Manual of Water Supply Practices

M3

Safety Management for Utilities

Eighth Edition

American Water Works Association

Manual of Water Supply Practices—M3, Eighth Edition

Safety Management for Utilities

Disclaimer
The authors, contributors, editors, and publisher do not assume responsibility for the validity of the content or any consequences of its use. In no event will AWWA be liable for direct, indirect, special, incidental, or consequential damages arising out of the use of information presented in this book. In particular, AWWA will not be responsible for any costs, including, but not limited to, those incurred as a result of lost revenue. In no event shall AWWA's liability exceed the amount paid for the purchase of this book.

If you find errors in this manual, please email books@awwa.org. Possible errata will be posted at www.awwa.org/resources-tools/resource.development.groups/manuals-program.aspx.

Senior Managing Editor/Project Manager: Melissa Valentine
Technical Editor: Suzanne Snyder
Technical Illustrations: Gillian Wink
Cover Composition: Michael Labruyere
Production: Innodata
Senior Specialist - Manuals: Molly Beach
Manuals Coordinator 2: Janet Greifinger

Library of Congress Cataloging-in-Publication Data
Names: Milligan, Frank (Hydraulic engineer), editor. | American Water Works Association, issuing body.
Title: Safety management for utilities / [edited] by Frank Milligan.
Other titles: Safety practice for water utilities. | AWWA manual ; M3.
Description: Eighth edition. | Denver, CO : American Water Works Association, [2022] | Series: AWWA manual ; M3 | Original title of earlier editions <3-4>: Safety practice for water utilities; <6-7>: Safety practices for water utilities. | Includes bibliographical references and index. | Summary: "This M3 update is the result of Health, Safety, and Environmental (HSE) Committee efforts to provide general but broad guidance to water providers as they develop health and safety programs. The audience for this manual is primarily utility management and other employees with assigned and overall safety responsibilities (e.g., utility safety managers, supervisors)"-- Provided by publisher. Identifiers: LCCN 2022010426 | ISBN 9781647170776 (paperback) | ISBN 9781613006115 (adobe pdf)
Subjects: LCSH: Waterworks--Safety measures--Handbooks, manuals, etc. | Water utilities--Employees--Health and hygiene--Handbooks, manuals, etc. | Water utilities--Safety measures--Handbooks, manuals, etc. | Handbooks and manuals.
Classification: LCC TD358 .S34 2022 | DDC 363.6/10684--dc23/eng/20220502
LC record available at https://lccn.loc.gov/2022010426

Printed in the United States of America

ISBN 978-1-64717-077-6 ISBN, electronic 978-1-61300-611-5

https://doi.org/10.12999/AWWA.M3ed8

American Water Works Association

American Water Works Association
6666 West Quincy Avenue
Denver, CO 80235-3098
awwa.org

Contents

List of Figures, v

List of Tables, vii

Preface, ix

Acknowledgments, xi

Chapter 1 **Establishing a Health, Safety, and Environmental Program 1**
Regulatory Requirements, 2
Regulatory Requirements and Best Practices for HSE Programs, 4
Developing an HSE Policy, 5
Management Leadership and Employee Participation, 5
Program Staffing and Structure, 9
Roles and Responsibilities, 10
Developing Written Programs, 12
Training and Education, 14
Record Keeping and Reporting, 16
Evaluation, Corrective Action, Metrics, and Continuous Improvement, 16
References, 18

Chapter 2 **Job Safety Analysis 21**
Advantages of a JSA, 22
JSA Development, 23
The JSA: An Important Safety Factor, 29
References, 30

Chapter 3 **Incident Management 31**
Incident Management Programs, 31
References and Bibliography, 41

Chapter 4 **Prevention Through Design 43**
What is Prevention Through Design?, 43
PtD Implementation in the Water Industry, 45
References and Bibliography, 51

Chapter 5 **Contractor Safety Management 53**
Making Safety Part of The Contract, 55
Contractor Prequalification and Selection, 58
Pre-Job Contractor Orientation And Planning, 61
Contractor Work Rules and Guidance, 62
Injury/Illness/Incident Reporting and Investigation, 63
Contractor Evaluation, 64
Third-Party Contractor Safety Management, 65
References and Bibliography, 66

Chapter 6 Common Hazards and Safety Programs **77**
Physical Hazards, 78
Biological Hazards, 95
Chemical Hazards, 98
Personal (or Health) Hazards, 107
Additional Information, 114

Appendix A Guidance Documents and Forms **115**

Appendix B Canadian Water Works Safety References **155**

Acronyms, 159

Index, 161

List of AWWA Manuals, 165

Figures

1-1 OSHA state plans, 3

1-2 Basic health and safety management program activities (plan, do, check, act), 4

1-3 Example of a large utility safety organization, 10

1-4 Example of a safety procedure header, 14

2-1 Identify potential hazards, 25

2-2 Hierarchy of controls: apply the highest level of control commensurate with the risk level, 27

2-3 Personal protective equipment, 28

3-1 Ratio of indirect to direct costs of on-the-job incidents, 32

3-2 Triangulation method, 35

3-3 Sketching motion through time, 36

3-4 Incident weed and root cause analysis, 37

5-1 Percent recordable cases with no lost workdays by type of operation (1999), 54

5-2 Percent recordable cases with lost workdays by type of operation (1999), 54

5-3 Percent of lost workdays by type of operation (1999), 55

A-1 CS-01 – Contractor pre-qualification information, 68

A-2 CS-02 – Contractor safety orientation checklist , 69

A-3 CS-03 – Facility visitor and contractor orientation/EHS requirements, 70

A-4 CS-04 – Contractor employee training certificate, 74

A-5 CS-05 – Contractor safety evaluation, 75

6-1 Confined space decision tree, 79

6-2 Hard, compact ground (left) can be braced at intervals, while saturated, filled, or unstable ground (right) needs additional sheeting or shielding to hold back loose soil., 81

6-3 Ground fault interrupter, 82

6-4 Proper lifting technique helps avoid back injuries, 85

6-5 Proper placement of cones, vehicles, pipes, and equipment help keep a field operation safe for workers, pedestrians, and public traffic, 91

6-6 Properly stacking and restraining pressure vessels reduces potential hazards, 100

A-6 Accident investigation program, 116

A-7 Accident/incident analysis, 119

A-8 Sample job description (light physical requirements), 120

A-9 Physical capacities requirements, 122

A-10 Early return to work program, 123

A-11 Hazard summary/action list, 127
A-12 Hot work permit, 128
A-13 Job safety analysis, 130
A-14 Sample job safety analysis, 131
A-15 Passing criteria, 133
A-16 Physical capacities requirements, 134
A-17 Release to return to work, 135
A-18 Pre-placement physical capacities test, 136
A-19 OSHA Quick Card, 137
A-20 Safety data sheet sample, 138
A-21 Temporary modified duty tasks, 148
A-22 Job description (heavy physical requirements), 149
A-23 Physical capacities requirements (heavy): maintenance and construction crew waterworks operator, 151

Tables

1-1 Example activities for observable management involvement in safety, 7

1-2 Example employee participation plan, 9

1-3 General utility HSE responsibilities, 11

1-4 Tracking of training, 15

1-5 Example table of incident action items, record keeping, and reporting, 17

2-1 Types of hazards to consider when conducting a JSA, 26

5-1 Examples of prequalification and selection information to be collected from the contractor, 59

5-2 Examples of prequalification and selection criteria to be used by the owner/utility, 60

A-1 Example Contractor Safety Program Process Forms, 67

This page intentionally blank.

Preface

The American Water Works Association (AWWA) recognizes the importance of protecting water industry professionals on the job as they work to deliver clean, safe drinking water to customers. As such, it created a safety policy which guides discussion of safety practices that are presented in this eighth edition M3 manual. The most current policy statement can be found on the AWWA website, https://www.awwa.org/Policy-Advocacy/AWWA-Policy-Statements/Safety.

This M3 update is the result of Health, Safety, and Environmental (HSE) Committee efforts to provide general but broad guidance to water providers as they develop health and safety programs. The audience for this manual is primarily utility management and other employees with assigned and overall safety responsibilities (e.g., utility safety managers, supervisors).

M3 cannot cover all situations and regulations that often differ by country, state, or province and therefore does not cite regional regulatory requirements (e.g., measurements, heights) but rather addresses US federal requirements such as those imposed by the US Occupational Safety and Health Administration (OSHA), with Canada-specific guidance included in the appendix.

M3 supplemental information and tools for use in developing HSE programs, including sample checklists, site inspection forms, programs, and job safety analyses, will be delivered electronically to those who purchase this manual through the AWWA website. They also are available at www.awwa.org/M3.

This page intentionally blank.

Acknowledgments

The AWWA Management and Leadership Division; the Health, Safety, and Environmental Committee; and the M3 Committee gratefully acknowledge the contributions made by the volunteers who drafted, edited, and provided the significant and critical commentary essential to developing M3. The Steering Committee members dedicated many hours in the final stages of preparation of this edition to ensure the overall technical quality, consistency, and accuracy of the manual.

Steering Committee Members

Frank Milligan, Committee Chair, Las Vegas Valley Water District, Las Vegas, Nev.

David Milligan, North Texas Municipal Water District, Wylie, Tex.

Jeff Deville, Watertown Municipal Utilities, Watertown, S.Dak.

Kenya Henderson, Las Vegas Valley Water District, Las Vegas, Nev.

Stephanie Passarelli, American Water Works Association, Denver, Colo.

Contributors to the Eighth Edition

Frank Milligan, Committee Chair, Las Vegas Valley Water District, Las Vegas, Nev.

Brian Exendine, Las Vegas Valley Water District, Las Vegas, Nev.

Dr. Nicholas Ashbolt, University of Alberta School of Public Health, Edmonton, Alberta, Canada

David Baird, III Applied EHS Management, Dallas, Tex.

Marissa Blackeye, Las Vegas Valley Water District, Las Vegas, Nev.

Mia Parton, Aeparmia Engineering, Austin, Tex.

Michael Jacobs, Tualatin Valley Water District, Beaverton, Ore.

Nick Kato, City of Gilbert, Gilbert, Ariz.

Peter Puglionesi, Applied EHS Management, Havertown, Pa.

Pete Wagner, SUEZ, Boise, Idaho.

This page intentionally blank.

Chapter 1

Establishing a Health, Safety, and Environmental Program

Pete Wagner, Brian Exendine

The safety and health of employees is a key concern in any organization. The goal of sending everyone home safely at the end of each shift should always be at the forefront of any organization's planning efforts. Staff knowledge and management support alone have proved effective in some small organizations but ineffective in others. A formal Health, Safety, and Environmental (HSE) program is a reliable and effective means of consistently protecting workers from occupational injuries and illnesses. Formal written programs can address specific hazards and prevention measures (e.g., fall prevention) and specific regulatory mandates that require a written program (e.g., hazard communication, respiratory protection). Furthermore, there is widespread agreement that written management systems improve organizational performance, including performance in occupational health and safety.

This manual serves to establish the key components that should be included in a formal HSE program. Chapter 2 describes job safety analysis (JSA) as a means to proactively examine and mitigate existing hazards on the worksite. Chapter 2 also introduces the hierarchy of controls, a methodology that is carried throughout the manual. A thorough HSE program will also outline the procedures necessary should an incident occur (incident management is discussed in Chapter 3). Recognizing that many hazards can be engineered out of a system, Chapter 4 addresses key aspects of prevention through design. A meaningful program should also be effectively communicated with contractors. Chapter 5 describes contractor safety management and some of the special hazards that accompany the construction phase. Last, Chapter 6 provides an index of common hazards. Many of these hazards should be considered in the development of any HSE program, and some can be evaluated depending on the type of system or its geographic location. In addition to planning for the hazards presented in Chapter 6, extreme storm events should be considered in the development of an HSE program. More information on safety during emergencies can be found in AWWA Manual of Water Supply Practices M19, Emergency Planning for Water Utilities.

The rest of this introductory chapter presents research that demonstrates the need for a formal HSE program. This chapter then describes how to establish an effective program by delineating management steps and development activities that should be applied to each of the components offered throughout this manual.

To emphasize the need for a formal HSE program, recent safety and health publications and the study noted below provide data that show companies with strong written safety and health programs that include key elements recognized by safety and health professionals and governmental agencies alike experience significant reductions in injury cases, claims, rates, and costs. According to V.J. Roberts (2020), OSHA data show that the average Voluntary Protection Program (VPP) worksites have lost, restricted, and/or transferred case rates that are 52 percent below the average days away/restricted time average for the same industry. According to a series of studies conducted in Canada on the effectiveness of Certificate of Recognition (COR) by Dr. C. McLeod and the Institute for Work & Health (2016), companies that implemented elements of the COR program have 12–14 percent greater reductions in lost-time injury rates than non-certified companies. Similarly, Dr. D. Kandilakis (2016) stated that companies in OHSA's Safety and Health Achievement Recognition Program (SHARP) whose written safety and health policies were reviewed over a 12-year period from 1999 to 2010, experienced an average 52 percent reduction in claims, an average claims cost decrease of 80 percent, and an average lost time per claim decrease of 87 percent.

REGULATORY REQUIREMENTS

The US Occupational Safety and Health Act of 1970 did not grant the Department of Labor's Occupational Safety and Health Administration (OSHA) authority over federal, local, or state government-operated facilities. In developing its program to authorize states to enforce federal regulations, OSHA recognized the need to encourage states to regulate state and local government workplaces. The process of attaining state authority involves submitting an acceptable state plan to OSHA, in which states are required by OSHA to enact legislation and regulations that are equivalent to federal rules and also apply to the state and local government workplace.

As a result, many states have their own OSHA rules that apply to state and local government-operated water treatment plants (WTPs), in particular states with OSHA plans approved by federal OSHA (i.e., OSHA plan states). In addition, some states have enacted equivalent rules that apply only to state and local government employees.

The federal government's employees are covered under Section 19 of the OSH Act. This has been augmented by Executive Order 12196, dated February 26, 1980, which requires agency heads to furnish employees a workplace free from recognized hazards that are causing or are likely to cause death or serious physical harm.

As a result, federal- or state-equivalent OSHA rules are applicable and enforceable if the facility is

- a privately-operated plant;
- operated by the federal government; or
- in one of the state's operating OSHA Plans (Figure 1-1).

For utilities in states listed as non-OSHA some governing body or legislation may be applicable, but no OSHA-equivalent safety and health standards exist. For example, in Texas public worker health and safety is managed by the Texas Department of Insurance, Workers Compensation Division, but there is no OSHA coverage of local municipal WTP workers. The Texas Workers' Compensation Act, section 411.101, states legislative policy of this division as "(1) promoting the adoption, application, and implementation of safety measures in industry and enterprise; (2) protecting workers against unsafe and hazardous working conditions; and (3) encouraging correction of any unsafe and hazardous working conditions in industry and enterprise." However, Texas and other non-plan states have no equivalent regulatory standards or enforcement, such as those provided by OSHA. So, for those organizations in non-plan states, the drivers for compliance and even excellence are not the same, and what limited data are available indicate that having a plan with standards and enforcement does make a difference.

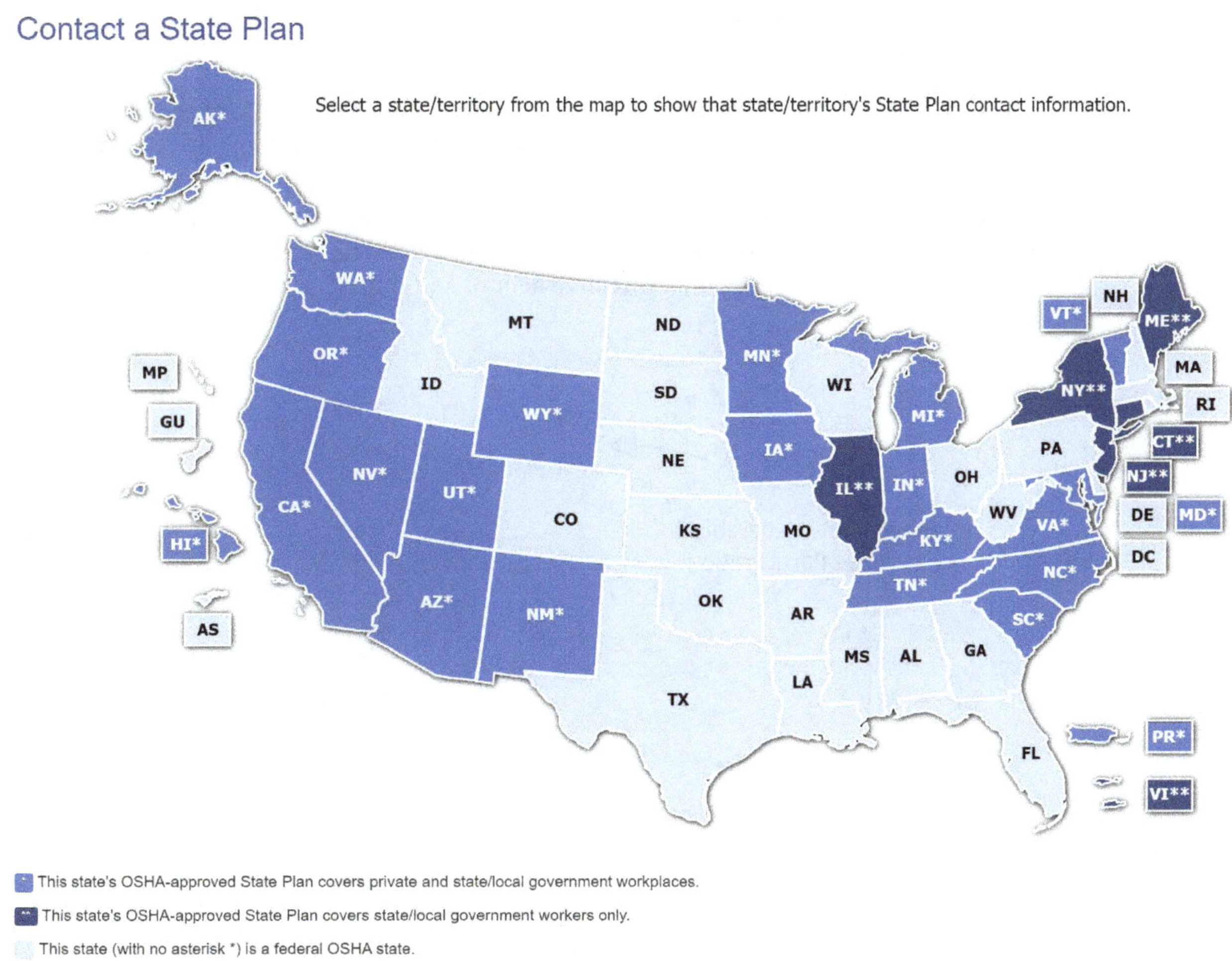

Figure 1-1 OSHA state plans

REGULATORY REQUIREMENTS AND BEST PRACTICES FOR HSE PROGRAMS

For those utilities that are regulated by OSHA's Process Safety Management (PSM) or the US Environmental Protection Agency (USEPA) Risk Management Program (RMP) regulations for control of hazardous chemicals, employee participation is a requirement that necessitates a written plan describing how employees are to be involved in implementation of those programs. This is critical for PSM and RMP because employees are working closest to the hazards and are in the best position to recognize and control those hazards. Likewise, employees such as operators and street crews can have the largest impact on injury prevention and safe work practices.

Injury and Illness Prevention Programs (IIPP) are written workplace safety management programs intended to improve workplace safety and health through good management practices and employee involvement. While some states, such as California, mandate that every employer implement an IIPP, there is no equivalent program requirement in the federal OSHA regulations, even though some specific regulations (e.g., confined space and lockout/tagout) call for written programs or plans. Regulations covering some industries or encouraging IIPP development exist in a total of 34 states. In recent years, OSHA has seriously contemplated such a requirement, and public comment on that proposal suggests that the regulation be based in part on one or both of two established industry consensus standards establishing best practices for IIPPs:

- ANSI/AIHA Z10-2012, Occupational Health and Safety Management Systems (American National Standards Institute)
- ISO 45001, Occupational Health and Safety
- OHSAS 18001-2007, Occupational Health and Safety Management Systems (Occupational Health and Safety Assessment Series Project Group) (ISO 45001 replaces 18001)

While terms may vary between sources, the basic concepts of an HSE management programs are management leadership and employee participation, planning, implementation, evaluation and corrective action, and management review, as illustrated in Figure 1-2.

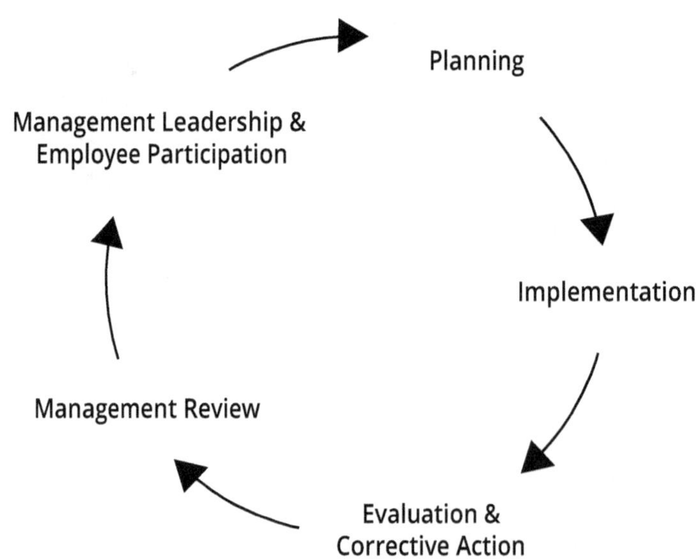

Figure 1-2 Basic health and safety management program activities (plan, do, check, act)

Based on these activities, this chapter describes the typical process for developing a HSE program and key attributes of successful programs. The primary topics addressed are

- developing a health, safety, and environmental policy;
- management leadership and employee participation;
- program staffing and structure;
- roles and responsibilities;
- developing written programs;
- training and education;
- record keeping and reporting; and
- evaluation, corrective action, metrics, and continuous improvement.

DEVELOPING AN HSE POLICY

"The quality of a person's life is in direct proportion to their commitment to excellence, regardless of their chosen field of endeavor."

—Vince Lombardi

With any endeavor, the quality of the outcome is in large part based on the level of commitment to that outcome. Management commitment is a key element that is universally understood as essential to safety excellence, but what is the actual expression of such commitment? Words? Resources? Presence? The short answer is "Yes" to all three, but it starts with words, initially presented through the HSE policy statement.

ANSI/AIHA standard Z10-2012 (Z10) states that a documented occupational health and safety policy is the foundation for any safety management system and that the policy must include the following four elements:

- Protection and continual improvement of employee health and safety
- Effective employee participation
- Conformance with the organization's health and safety requirements
- Compliance with applicable laws and regulations

Z10 states the policy must be dated and signed or otherwise officially authorized and endorsed by top management and made available and communicated to employees. It is also helpful to share this information with other utility stakeholders and affected par ties such as contractors, customers, and vendors.

A brief, written safety policy statement might be worded as follows:

The [Anytown Water Utility] highly values the safety of its employees and the public. We are committed through the joint efforts of every employee (including the directors, managers, supervisors, and crews) to keep our workplace free from health and safety hazards and to prevent injuries and illnesses. We continually evaluate our safety performance and make improvements where necessary to maintain a safe workplace. Through routine and effective training and diligent implementation of our safety program, including all policies and procedures of [Anytown Water Utility], we will maintain the safety of our employees and full compliance with all applicable Federal, State and Local regulations.

MANAGEMENT LEADERSHIP AND EMPLOYEE PARTICIPATION

Management commitment is one of the most important requirements for success in any safety performance program. In order to overcome unsafe practices that have become status quo and to be heard above the noise of day-to-day activities and priorities, management commitment must be expressed in meaningful and ongoing ways. A written policy is generally not enough to achieve stated safety goals. Compliance requires the active engagement of all employees to develop a culture aligned with a commitment to safety.

If a utility historically has not used a written safety management program or has had some limited level of safety management but desires to move to the next level, there may be some resistance to this effort. There may be a historic organizational culture that has not placed a priority on regulatory compliance. New managers and supervisors may have history or experience with different safety cultures and may be implementing those cultures at their new workplace. While these individuals may want to do the right thing with regard to safety issues, their previous experience could have had inadequate safety staffing levels, a lackadaisical approach to safety, or other significant obstacles to a successful safety program. In that environment, these employees may have accepted and internalized historical organizational values that had assigned a lower priority to safety compliance, or they may doubt the organization will provide the necessary support (financial, manpower, training) needed to comply.

Human nature can make it difficult for employees to accept new approaches at face value when the new methods appear to conflict with years of experience. Employees may respond negatively to being given safety responsibilities unless there is clear communication that safety is a co-priority with meeting production, repair, construction, and financial objectives and there is sufficient technical safety support and oversight. Management must regularly reinforce the "our job is safety compliance" message and provide tangible evidence of its commitment through regular employee engagement and by providing the resources for achieving the safety objectives.

While policies and management plans are effective in setting the stage for a company commitment to safety, these plans and policies are best reinforced through management's active participation. Examples include the following:

- Directors, managers, and supervisors integrate regular and active involvement in safety compliance efforts into their day-to-day business (Table 1-1).
- Directors and managers discuss safety goals, their commitment, and employee accountability not only with their direct reports, but with other employees.
- Safety objectives and accountability are integrated into performance evaluation goals for directors, managers, supervisors, and all employees.
- Goals are set that measure poor performance by noncompliance rather than by the occurrence of an accident. Goals should be pro-active and specific. Examples could include conducting site visits, audits, reporting and correcting unsafe conditions, etc.
- Corrective and disciplinary action is taken when managers, supervisors, or employees knowingly and repeatedly fail to meet their safety obligations. This is necessary to demonstrate that the organization is serious about meeting its compliance obligations.
- Employees who excel at EHS, just like any other tasks, need to be recognized.
- The safety budget process is openly defined and communicated, and adequate funding is budgeted for safety staff, equipment, and supplies, including unanticipated contingencies such as addressing noncompliance issues quickly when they arise.
- A mechanism is available to expedite funding, engineering, and/or procurement of critical safety needs.
- Currently known critical safety needs are addressed as soon as possible.

Table 1-1 Example activities for observable management involvement in safety

	Routinely	Weekly	Monthly	Quarterly
Director	Demonstrate leadership, show visible support Participate in incident investigation process	Site safety visit: provide a visible presence in work area with a safety emphasis Participate in weekly safety meetings	Review safety performance reports Hold team members accountable for safety requirements	Conduct a safety observation Participate in audits/ inspections Lead safety toolbox meetings Participate in reward/ recognition event
Managers	Demonstrate a personal commitment to safety at all times Participate in incident investigation process	Lead weekly safety meetings	Conduct safety observations Participate in audits/ inspections	
Supervisors	Continuously practice, educate, and involve employees in hazard identification and use of safe work practices Investigate all incidents and near misses Recognize safety achievements Consistently implement disciplinary program	Lead daily pre-job safety briefings that review activities, hazards, and controls Lead weekly toolbox safety meetings	Participate in audits/ inspections	

Adapted from Clark 2008

Responding quickly to critical resource needs conveys management commitment to safety. Measures such as a priority approval process for safety critical needs and/ or a contingency budget reserve are necessary to give management the ability to take actions that will bolster credibility and demonstrate management commitment. Table 1-1 provides additional specific criteria for management involvement that can be incorporated into performance requirements for each member of management and is observable by employees.

Achieving a Culture of Excellence

Many organizations look at compliance as the goal for their safety programs but remember that regulations are often minimum standards and just achieving compliance may not meet all safety goals for a given utility. Moving beyond compliance to safety excellence requires the active engagement of all employees to develop a culture aligned with the newly defined commitment. Culture can be one of the most difficult things to change. Culture change efforts in some larger organizations show that only a small fraction of management and employees will embrace new safety objectives and programs at their inception. The majority will adopt a wait-and-see attitude, and a small fraction will actively resist it. Success requires years of persistent communication, training, and—most important—showing this commitment in tangible ways.

It is never too late to start, and each utility at any time has the opportunity to establish and convey its safety priorities and establish a positive safety culture. Communication and employee participation as outlined in the following sections can improve the chances of success.

Communication

The following measures are suggested to actively, continually, and effectively communicate safety policy and programs to employees, supervisors, and managers:

- Develop or improve existing model safety tools and plans to facilitate meeting basic safety compliance requirements.
- Create a safety communication vehicle (newsletter or section in an existing utility communication) to regularly communicate to employees on safety policies, programs, new developments, performance, and positive examples.
- Encourage active communication, cooperation, and coordination between individuals with corporate- or management-level safety functions and the supervisors and field staff with safety responsibilities through a safety coordination group that meets regularly.
- Set up an intranet safety page with corporate policies, procedures, model programs, and links to outside resources (e.g., federal and state regulations).
- Adopt incentive programs to promote active participation in safety compliance and excellence. While cost may be an issue, there will be financial returns in reduced injury and environmental costs. Measures should certainly include no-cost and low-cost measures such as featuring "Safety Employee of the Month/Year" in the safety publication and recognition at employee events. Consider low-cost awareness programs, such as caps or shirts for reaching milestones or performance goals as well as more sophisticated moderate-cost programs.

Employee participation. As important as management commitment and its impact on effective safety program implementation and performance is employee involvement.

The intent of employee participation is to encourage employees to

- have meaningful involvement in the structure, operation, and pursuit of the objectives of the safety program;
- identify tasks, hazards, and risks, and possible control measures; and
- participate in implementation of control measures.

In practical terms, employee participation in an effective safety program means pro viding a role for supervisors, hourly employees, and employee representatives in activities such as procedure development, training development, job safety analysis, incident investigations, health- and safety-related audits, and all aspects of the safety system planning and development process. Probably the most common mechanism for employee participation is participation in health and safety committees (adapted from ANSI/AIHA Z10-2012).

A safety committee is made up of both employee and management representatives. Typical duties include

- workplace self-inspections;
- accident investigations;
- developing safe work practices;
- developing written safety programs; and
- facilitating safety training.

The committee also can help promote other activities that encourage employees to support the organization's safety program. Table 1-2 provides an example of defined employee participation for a few sample safety compliance areas.

Table 1-2 Example employee participation plan

Safety Program Element	Employee Involvement	Designated Employee(s)
Safety training	Supervisors, operators, and field crews will participate in weekly safety talks about various safety topics. Selected staff may receive more intensive training on particular subjects related to their work situations.	All employees
Incident investigations	Involved personnel will cooperate with investigation team. Selected personnel may participate in the investigation as requested by the incident investigation team lead.	All employees
Emergency planning and response	All employees and contractors currently on-site will be provided emergency warning and evacuation training. Trained employees will be on emergency response team. Emergency response team members will assist in reviewing the emergency response plan.	All employees and affected contractors Emergency response team members
Safety audits	Selected personnel may participate in audits and cooperate when interviewed by auditors. All employees will be provided emergency warning and evacuation training.	All employees

PROGRAM STAFFING AND STRUCTURE

Safety staffing demands are not solely based on an organization's employee count. There is a baseline of activity and expertise needed for the organization-wide and facility levels that is independent of number of facilities and a component of effort that is dependent on the number of employees/facilities. There are certain basic good practices and principles that can apply to any safety management system, including the following:

- Having a safety-accountable executive-level position makes it easier for organizations to achieve and maintain a high level of performance.
- All directors and facility managers must be accountable for safety compliance, support safety efforts for their operations, and regularly reinforce the importance of safety compliance to their employees.
- In large organizations, a central safety group can operate on a high level, providing policies and programs, assessing and auditing performance, encouraging coordination, and providing efficient shared services (e.g., reviewing and communicating emerging regulations).
- Safety compliance is more effective when experienced, trained safety staff provide close support to operations.
- Larger facilities or clusters of facilities may have a safety workload that justifies a designated safety specialist. If these positions report to operations, they should work closely with corporate safety groups (i.e., a "dotted-line" relationship) and follow corporate policies and programs.
- Safety staff at multiple levels within an organization must have clearly defined roles and responsibilities to achieve common safety goals and avoid duplication of effort, divergent programs, and potentially adversarial relationships.
- Everyone in the organization should have some responsibility for supporting the safety programs (integrated into their performance goals).

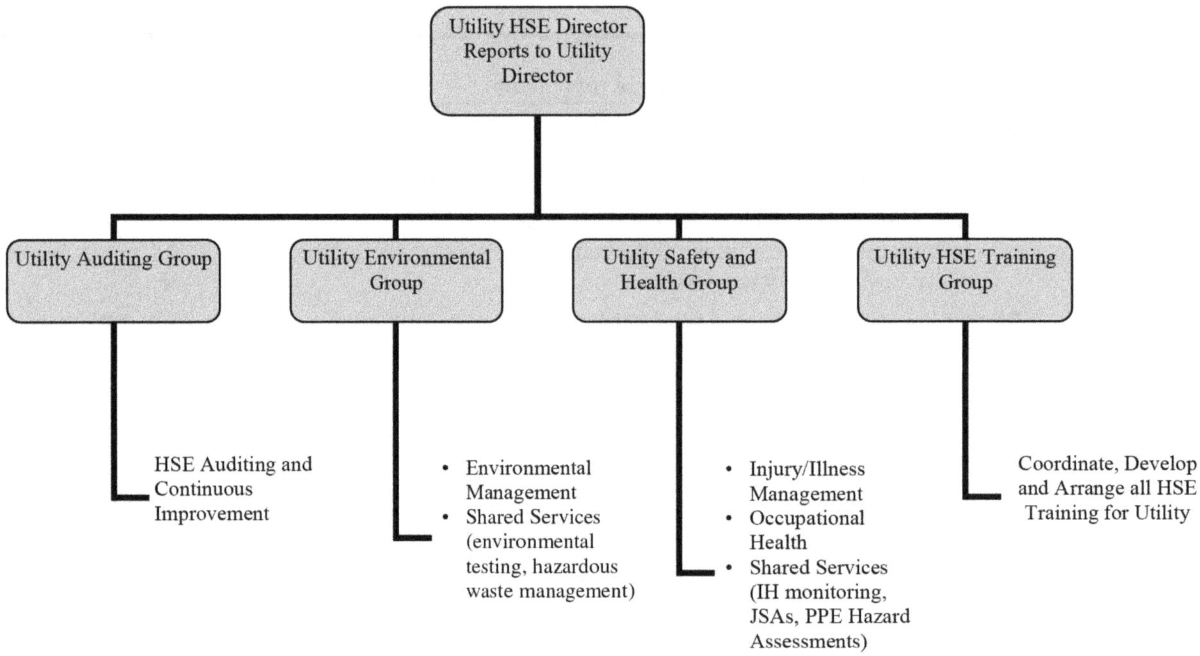

Figure 1-3 Example of a large utility safety organization

The staffing requirements often change and become more clearly defined as programs mature. A higher level of staffing may be necessary in the first few years of new safety program implementation because of time-intensive tasks such as program development and initial training. Once the programs are developed or revised and all initial training is completed, it is typical for a utility to reassess staffing levels based on a regular maintenance level of effort (e.g., refresher training and program support).

Figure 1-3 represents how the safety staff might be organized at a large utility with several facilities.

Smaller utility staffing. In smaller utilities the responsibility for managing safety may fall to a single safety manager who is supported by a training coordinator and/or administrative assistant. This can be challenging because, even though the department is smaller, many of the requirements and responsibilities are the same. A safety manager of a smaller utility is often required to not only develop the HSE program but also perform daily tasks such as inspections. Managers of small systems can find assistance from their workers' compensation insurance carrier's loss control department. OSHA also offers consulting services that small utilities can use as a source of expert advice.

Staff qualifications. Qualified staff is critical to safety success; therefore, when hiring and promoting safety staff, it is critical to select highly qualified individuals with the necessary training and education to fill as many safety needs as possible. If a small utility has short-term assignments or non-routine needs calling for skills and manpower in areas such as industrial hygiene and environmental management systems, outside resources can be contracted if necessary.

ROLES AND RESPONSIBILITIES

Whether regulated by OSHA or not, some basic activities must be performed by any organization in order to provide a safe working environment for its employees, including job safety analysis, hazard prevention and control, and training. Beyond this, more detailed and administrative functions must be performed, such as

- injury/illness recordkeeping and reporting;
- workers compensation cases;
- injury case management;
- incident investigation;
- insurance inspections/requirements;
- employee health and safety concerns; and
- union/labor safety meetings and issues.

Often these responsibilities are spread across the organization (especially in smaller utilities) with no one person having overall responsibility for ensuring that these duties are completed. To ensure that all required and desired activities are being performed consistently and effectively, each utility should define the roles and responsibilities for HSE compliance. Table 1-3 provides a general distribution of HSE responsibilities that can be adapted to any organization, whether with dedicated safety staff or only production or administrative staff with dual responsibilities.

In smaller organizations with less staff, time must be spent on programs and tools that can quickly and efficiently be deployed to ensure that safety actions are completed and not continually put off in favor of seemingly more pressing production or financial demands. Regardless of the size of the organization, successful implementation of a safety program depends on each responsibility being formally assigned and effectively executed.

It is essential that the staff at every level make every effort to function as one team and collaborate on procedure and program development and assist each other in implementation, training, audits, and other safety program elements.

Table 1-3 General utility HSE responsibilities

Corporate/Utility level	Regional/Divisional Level	Facility Level
Develop policies.	Implement policies.	Implement programs and procedures.
Coordinate development of utility-wide programs and procedures.	Participate in development of utility-wide programs and procedures and develop tailored procedures, if necessary.	Participate in regional/divisional HSE committees to identify needs and enhance facility HSE capabilities.
Track emerging laws and regulations that affect utility operations and communicate to the regions/ facilities.	Implement HSE programs throughout the region/division by direct support, training, and oversight, as appropriate.	Regularly communicate HSE procedures and precautions to employees.
Provide shared services to the regions/ facilities where it is more efficient to do so at the corporate level (e.g., training, specialty services).	Provide expert services to the facilities where it is more efficient to do so at the utility or regional level (e.g., industrial hygiene, hazmat).	Conduct regular reviews of potentially hazardous operations to verify that proper protections and procedures are used.
Conduct audits of regional/facility HSE compliance and assessments of program implementation efforts.	Conduct regular facility reviews to assess hazards and identify the need for further compliance assistance.	Complete required inspections, training, and records.

DEVELOPING WRITTEN PROGRAMS

The process of establishing a written health, safety, and environmental program will help identify what has to be done in an organization to control hazards and promote greater health and safety awareness among employees, supervisors, and all key stakeholders including contractors, vendors, and customers. Developing a written program allows utilities to outline the policies and procedures that are necessary to achieve stated injury and illness prevention goals. Two levels of written documents will help define and guide a safety program:

- Management programs/procedures (e.g., program improvement [audits], incident investigation, record keeping)
- Compliance programs/procedures (e.g., lockout/tagout, hazard communication, confined-space entry)

Management program development. The purpose of a management program is to define the roles and responsibilities of individuals as well as the process mechanisms needed to implement and maintain an effective HSE program, resulting in a systematic, organized, and comprehensive approach to managing and achieving both regulatory compliance and excellence in health and safety performance. While the development of written programs begins within the planning activities, the written document itself needs to specifically address each of the basic elements of safety management: management leadership and employee participation, planning, implementation, evaluation, and corrective action. The following management system components are examples of the types of procedures used to plan, implement, and evaluate the safety and health program.

HSE responsibilities. Each organization is responsible for having in place clearly defined procedures to comply with health, safety, and environmental regulations. Included within this procedure should be assigned roles and responsibilities for those individuals who are responsible for implementing and maintaining an effective health and safety program.

Management program compliance. The compliance procedure can be used to ensure that all applicable regulatory requirements are encompassed in the utility's program documents by correlating the regulatory standards to the contents of the implementation programs.

Employee participation/awareness. This component includes methods for communicating with employees—in a manner readily understood by all affected employees—on matters relating to safety and health. This includes provisions designed to encourage employees to inform the employer of hazards or concerns at the worksite without fear of reprisal. It is also intended to ensure that employees are provided with full awareness of their rights and responsibilities under the utility program.

Employee training. Training is one of the most important elements of any HSE program. It allows employees to learn their job properly, brings new ideas into the workplace, reinforces existing ideas and practices, and puts the program into action. Benefits from safety training include fewer work-related injuries and illnesses, minimizing risks to the environment, and reduced employee stress and worry caused by exposure to hazards. At a minimum, the safety training should include all applicable regulatory man dated training for both supervisors and employees. This program may be merged with or implemented in conjunction with programs that manage and track other required training for water treatment operators.

Injury/incident investigation and reporting. This document contains the basic procedures and requirements for recording and reporting occupational illnesses and injuries as well as identifying the cause(s) of the accident or near-miss occurrence. In turn, this information can be used to implement corrective changes to preclude the recurrence of similar incidents.

Program improvement (auditing). An important part of any HSE program includes self-assessment and compliance measurements. By defining and identifying measurable performance marks and auditing against those marks, utilities are able to define specific goals and achieve continuous HSE program improvement resulting in fewer occupational injuries and illnesses.

These are some basic management procedures that should be part of any written HSE management program, and these can be expanded with additional procedures to address any topics that help those with safety responsibility understand and consistently plan, implement, and continuously improve performance. Management programs should be written in conjunction with compliance programs that ensure the observance of HSE standards.

Compliance programs. Compliance programs and procedures detail the procedures, practices, and instructions for the day-to-day implementation of the health and safety systems in accordance with regulatory requirements and your utility policy. These can be augmented by a safe work practices manual that serves as a quick pocket reference intended to supplement written procedures and training. The pocket manual provides guidance in the absence of any written procedure to assist employees and con tractors in the safe performance of their duties. OSHA's general industry standards are broken down into several subject matter areas. Some of these require written compliance programs, such as

- bloodborne pathogens;
- confined space;
- control of hazardous energy;
- hazard communication;
- laboratory chemical hygiene;
- hearing conservation; and
- respiratory protection.

Other compliance areas do not require written programs but do require specific documentation. Developing formal written procedures for addressing these other issues is beneficial to ensure consistent compliance and implementation. One example of this is in personal protective equipment (PPE) where no written program is required but, under 29 CFR 1910.132 (2004), there must be a written and "certified" assessment of the workplace hazards.

Procedure development. People often learn, or are reminded, how to perform a task through procedures, and it is highly likely that most people will forget how to do tasks that are not repeated with frequency, hence the need for written procedures. Additionally, procedures often include checklists and forms that provide extra control and documentation to help ensure that work is performed properly and in accordance with established methods and safety precautions. Lack of procedures or a failure to follow procedures is a key root cause in many incidents and near misses both large and small.

Written procedures should not be too long or written as though only experts will be reading and using them. Here are some basic guidelines for writing effective safety procedures.

Know your objectives. Developing effective safety and health procedures requires a clear idea of what the procedure is to accomplish. Some objectives may be to provide a basis for discipline when safety rules are ignored, establish safe work practices for certain tasks, or to define how your organization will comply with safety regulations.

Know your audience. Very often safety procedures are written for safety professionals, supervisors, trainers, and other members of management who will interpret and enforce company regulations on safety and health (e.g., respiratory protection). Some

Anytown utilities	SAFETY MANUAL	
Number: SP-1	Subject: Lockout/Tagout Program	
Page 1 of 30	Current issue: Rev. 1	Issue date: 04 Dec 12
Approved for use by:	I.M. manager	

Figure 1-4 Example of a safety procedure header

procedures, however, will have an audience of employees or contractors who will follow the rules (e.g., excavation and trenching or emergency response). Understanding who will be using the procedures and in what environment can help determine what writing and layout style (e.g., significant white space and bold type for emergency situations) will best suit the audience.

Use a consistent format. The format used should make it easy to find a specific rule, procedure, checklist, or form when a safety or health question comes up. Consider using electronic documents for easy access by employees in the office or in the field. Additionally, on each procedure, consider using a consistent header, such as the one in Figure 1-4, that includes

- utility name or logo;
- the manual name (if you have more than one manual);
- section number (if more than one section);
- procedure number and a brief procedure title;
- date of issue and revision number;
- page number (as "page 1 of 3"); and
- optional approval signatures/initials and dates.

Implementation matrix. Even with concise and well-written programs and procedures, compliance is still a challenge and it can be difficult to keep up with all the requirements outlined in the procedures. Limited staffing in smaller organizations can mean that some individuals may be assigned safety compliance responsibilities in addition to other human resources, operations, or management duties. Regardless of the level of staffing, understanding all the required tasks for compliance can be overwhelming. Required actions may include the one-time activities recommended to develop and implement complete safety programs as well as those recurring activities required to meet ongoing federal and state health and safety regulatory requirements.

One tool that can help manage these obligations is an HSE matrix or calendar of regular requirements that can be used by facility personnel with compliance responsibility to schedule and complete the implementation activities. This matrix should include the required initial and refresher training (see next section) and record-keeping requirements. The matrix should also include a column of key actions to achieve and maintain compliance. For recurring activities, a date can be added to track when those obligations are due. Alternately, there are many electronic training systems/databases that track all employee training, not only EHS training.

TRAINING AND EDUCATION

Many standards promulgated by OSHA explicitly require the employer to train employees in the safety and health aspects of their jobs. Other OSHA standards make it the employer's responsibility to limit certain job assignments to employees who are "certified," "competent," or "qualified"—meaning that they have had special previous training, in or out of the workplace. Designated personnel are selected or assigned by the employer or the employer's representative as being qualified to perform specific duties. These requirements reflect OSHA's belief that training is an essential part of every employer's HSE program for protecting workers from injuries and illnesses. Many researchers conclude that those who are new on the job have a higher rate of accidents and injuries than more experienced workers. Increasing knowledge of specific job hazards and proper work practices to lower injury rates is the primary goal of training employees. Training in the proper performance of a job is time and money well spent, and utilities should regard it as an investment rather than an expense. An effective program of health and safety training for workers can result in fewer injuries and illnesses, better morale, and lower insurance premiums, among other benefits. OSHA Publication 2254 (2015) provides the following key steps for developing an effective training program:

- Determine if training is needed
- Identify training needs
- Identify goals and objectives
- Develop learning activities
- Conduct the training
- Evaluate program effectiveness
- Improve the program

Tracking Training

Each employee's training requirements will be commensurate with the hazards the individual may be exposed to in work assignments. The training modules should be developed, and training should be assigned based on job assignments (e.g., confined space entrant for someone required to enter a tank) or workgroup (e.g., all operators or all emergency response team members). In any case a simple spreadsheet or matrix similar to that shown in Table 1-4 can be used to document required training.

Training Documentation

All training needs to be documented. The best training records involve a package of information, including items such as the following:

- The date of the training
- The instructor or presenter
- The content of the training (include slides, handouts, or agenda)
- The training method (e.g., classroom, on the job, tailgate)

Table 1-4 Tracking of training

		Personnel Required to Take Course		
Course/Module	**Frequency**	**Operations**	**Maintenance**	**Office/Admin**
Lockout/tagout (LO/TO)	Initial	√	√	
LO/TO refresher	As required*	√	√	
Hot work	Initial		√	
Hazard communication	Annual	√	√	√

* OSHA regulations require retraining in certain situations (e.g., unsafe work practices or change in equipment).

- A sign-in sheet of attendees with signature required
- If applicable, the means used to verify the training was understood (e.g., test scores)

These records can provide evidence of good faith and compliance with OSHA standards and support audits of the program. It can also help answer one of the first questions an accident investigator will ask: "Was the injured employee trained to do the job?"

RECORD KEEPING AND REPORTING

Accurate and consistent record keeping is a key factor in compliance and excellence.

Therefore, maintaining up-to-date documentation and records is critical for complying with specific regulatory requirements, where required (e.g., OSHA 300 logs), but also for ensuring that the safety management program can be adequately understood, effectively evaluated, efficiently operated, and systematically improved. As part of the management system procedures discussed above, it is helpful to establish, implement, and maintain a procedure(s) for the identification, storage, protection, retrieval, retention, and disposal of records. This procedure can help ensure that all required records and reports of occupational illnesses and injuries are maintained in accordance with regulatory requirements (e.g., OSHA 29 CFR 1904). Another purpose of this procedure is to prevent recurrences of occupational injuries and illnesses by documenting and investigating major incidents, identifying root causes, and taking corrective actions. Incident investigation helps to identify hazardous areas and work assignments, systemic or repetitive patterns, and improvements that will reduce the overall injury and illness rate.

This procedure should include necessary and useful record-keeping forms or links to forms (OSHA and/or state equivalent forms), such as the following:

- Log of Work-Related Injuries and Illnesses (OSHA 300)
- Summary of Work-Related Injuries and Illnesses (OSHA 300A)
- Injury and Illness Incident Report (OSHA 301)
- Utility Incident Investigation Report (IIR) (unless a separate IIR procedure exists)
- Instructions for Recording and Reporting Injuries and Illnesses

Another helpful record-keeping tool is a summary of incident action items to aid in completing the correct records for specific incidents. Table 1-5 provides an example of a table that can be customized to meet specific utility policies as well as federal and state requirements for record keeping and reporting. This table includes some regulatory requirements, such as reporting to OSHA on fatalities, but it can be modified to meet individual expectations with regard to reporting.

Ultimately, the data you record need be uniform and accurate to assure the consistency and validity of the statistical data reported to OSHA. Accurate records are used for many purposes, including audits, performance measurement, program improvement, and resource allocation to critical operations. This information can also help utilities analyze their health, safety, and environment and make improvements where necessary.

EVALUATION, CORRECTIVE ACTION, METRICS, AND CONTINUOUS IMPROVEMENT

One of the primary goals of any performance improvement program is to ensure that the organization is in full regulatory compliance with existing applicable federal, state, and local HSE regulations. A secondary objective is to use performance improvement

Table 1-5 Example table of incident action items, record keeping, and reporting

Severity	Incident Threshold	OSHA 300 Report	Incident Investigation Report (IIR)	Internal Notification	Report to OSHA within 8 hours
Major	Fatality of an employee or third party on utility premises, in a utility vehicle, or on utility business	Yes	Yes	Yes	Yes
	Hospital admission of three or more employees	Yes	Yes	Yes	Yes
	Hospital admission >24 hours of any employee	Yes	Yes	Yes	No
Moderate	Days away from work (lost workday case)	Yes	No	Yes	No
	Medical treatment	Yes	No	Yes	No
Minor	First aid case	No	No	Yes	No
	Minor property damage/ downtime resulting in < $10,000 damage s	No	No	Yes	No
	Near miss	No	No	Yes	No

processes, such as audits, as an opportunity to informally train personnel in compliance requirements and procedures and to raise awareness regarding utility HSE goals. Ultimately, a performance improvement program should be more broadly focused than just maintaining strict compliance; it should also include determining whether the established policies and procedures have been properly implemented and maintained, assessing measures to promote HSE excellence, identifying underlying root causes of HSE noncompliance incidents, and conducting evaluations of HSE management systems.

Measuring and monitoring are key elements of any safety program. Measuring and monitoring are used to determine whether the program is working as designed, any deficiencies are corrected, and opportunities for improvement are rolled back into the program.

Among other methods, Zl0 lists the following methods to monitor, evaluate, and control hazards:

- Workplace inspections and testing
- Exposure assessment
- Injury, illness, and incident tracking
- Employee input
- Occupational health assessments

The organization can use these measures to identify and correct problems and identify opportunities for risk reduction before injuries or illnesses occur.

Other main components of the continuous improvement and feedback process are

- incident investigation;
- audits;
- corrective and preventative action; and
- feedback to the planning process.

Incident investigation. Incidents and even near-misses can be indicators of potential problems and as such should be thoroughly investigated to determine root causes and identify corrective actions.

Audits. Audits of both the management system and regulatory compliance are valuable and a necessary part of the safety program. The system audit should determine if the management system procedures and their implementation are effective in managing the program. Compliance audits and inspections should outline appropriate corrective and preventive action activities as well as identify areas for feedback to the planning process for improving the management system. When possible, audits should be conducted by individuals independent of the activities being examined. This does not necessarily mean individuals external to the organization, but the auditors should be independent of the activities being evaluated.

Corrective and preventive action. As part of its safety management system, each utility should establish and implement procedures for identifying and investigating noncompliance issues, developing effective corrective and preventive actions, and tracking actions to closure. An effective procedure of this nature should include, at a minimum, the following elements:

- A method to address and investigate potential or actual non-compliance or safety issues
- A system to assign responsibility and provide resources to initiate and complete corrective or preventive actions
- A process to prioritize corrective or preventive actions based on the significance of risk and include deadlines for completion
- A process to track corrective actions to closure and communicate status to leadership and affected personnel
- A quality assurance process to confirm that corrective or preventive actions are adequate and appropriate to address the non-conformance
- A method to follow up on completed corrective actions to confirm that the intended results were achieved

Feedback to planning. As described in Z10, the findings and lessons arising from all the evaluation and corrective action activities must become part of the information that feeds back to the employee participation process, ongoing planning process, and management review. This information is used to help determine the underlying causes and other factors contributing to system or risk control failures, and hence is used to establish revised objectives and implementation plans. This feedback loop is an essential component of the continual improvement process.

REFERENCES

ANSI/AIHA Z10. 2012. American National Standard for Occupational Health and Safety Management Systems. American Industrial Hygiene Association.

Clark, J. 2008. Developing a Zero Incident Safety Culture, A Case Study. American Society of Safety Engineers Professional Development Conference and Exhibition, Las Vegas, Nev.

Institute for Work & Health. Employers Certified by COR Programs Have Greater Reduction in Injury Rates: Studies (2016).

ISO 45001, Occupational health and safety management systems — Requirements with guidance for use. International Organization for Standardization.

Kandilakis, D. 2016. Evidence That Injury Prevention Programs Are Cost Effective for Small Business Too! LinkedIn.

Occupational Safety and Health Administration. 2019 (6th ed.) *OSHA Construction Safety Handbook*, Neenah, Wisc.: J.J. Keller & Associates Inc.

Occupational Safety and Health Administration. 2004. *Code of Federal Regulations.* Title 29, Subpart I, §1910.132, Personal Protective Equipment.

Occupational Safety and Health Administration. *Code of Federal Regulations*. Title 29, §1904.

OSHA Publication 2254. 2015. Training Requirements in OSHA Standards and Training Guidelines (revised). Occupational Safety and Health Administration.

Roberts, V.J. How-To Guide to OSHA's Voluntary Protection Programs. 2020. [Online.] Accessed May 10, 2022.

US Department of Labor, Office of Inspector General, Office of Audit. 2000. Evaluating the Status of Occupational Safety and Health Coverage of State and Local Government Workers in Federal OSHA States (report 05-00-0001-10-001.) [Online.] Accessed May 10, 2022.

To access the 2000 report directly above, visit:

https://www.oig.dol.gov/public/reports/oa/2000/05-00-001-10-001.pdf

https://www.oig.dol.gov/cgi-bin/oa_rpts-v6.cgi?s=&y=2000&a=all (Scroll down to no. 48.)

This page intentionally blank.

Chapter 2

Job Safety Analysis

Frank Milligan, Brian Exendine

Workers are injured or killed at workplaces across North America every day. Accidents can mean both human loss and financial loss to the employer and the employee. How can these losses be minimized? The best way to prevent losses caused by accidents and occupational illness is to develop a solid, working program for accident prevention. Systematically looking at workplace operations, establishing proper job procedures, and ensuring all employees are properly trained can help prevent workplace injuries and illnesses.

As a part of an effective health, safety, and environmental (HSE) program, a job safety analysis (JSA) has proved to be an effective accident and occupational illness prevention tool in many industries. The JSA, also known as a job hazard analysis (JHA), is a safety management tool in which the risks or hazards of a specific job in the workplace are identified and measures to eliminate or control those hazards are determined and implemented. Most safety programs are considered reactive, an action in response to past incidents. A JSA is considered a proactive approach to workplace safety. A JSA is a process of systematically evaluating certain jobs, identifying the hazards or potential hazard associated with each step of the job, and finding effective control measures to eliminate or reduce the risk or hazards and make the workplace as safe as possible before those hazards have a chance to become accidents.

The Occupational Safety and Health Administration (OSHA) defines a JSA as a means of "carefully studying and recording each step of a job, identifying existing or potential job hazards, and determining the best way to perform the job to reduce or eliminate these hazards."

JSAs are a valuable tool for water utilities that have a commitment to reduce and prevent accidents and illnesses on the job. Developing and implementing safe job procedures and training employees to work properly to prevent accidents is likely to result in fewer worker injuries and illnesses, reduced workers' legal claims, more effective work methods, and increased productivity.

JSAs play an important role in effectively managing HSE programs. This chapter provides a guideline for how JSAs can be used, prepared, and implemented. Key terms for understanding the function of a JSA are as follows.

A *hazard* is a potential for harm. If left uncontrolled, a hazard could result in injury or harm. A hazard can be a physical object, chemical, noise, radiation, extreme heat or cold, electrical energy, or anything else that has the potential to cause harm.

An *accident* is an unintended occurrence that may result in injury, loss, or damage. As soon as people are involved, a hazard may cause an accident. Someone slipping in an oil spill would be an accident.

An *injury* is the result of an accident. A sprained wrist or broken arm from a fall would be an injury.

A hammer balanced on a windowsill is a hazard. Knocking the hammer off the sill is an accident. The hammer striking someone and causing a cut or laceration is an injury.

ADVANTAGES OF A JSA

The JSA is used to determine physical, procedural, and/or environmental hazards that do or could exist. JSAs also identify actions of personnel that could result in accidents or injuries and preventive measurements which would eliminate or control hazards, so a job can be performed safely.

The JSA not only helps supervisors and employees become aware of hazards on the job, but it also has many other uses as well. A JSA may be used

- to determine specific training or skill development an employee should receive;
- as a basis for inspection;
- as an informational tool for accident investigations;
- as part of a continuing communication program on employee safety and health awareness; and
- as a documented hazard assessment for PPE selection.

A JSA is especially helpful as a training tool because this tool provides an organized system for training both new and existing employees. Once a JSA has been prepared for a specific job, training new employees on that job will become more consistent. The entire job procedure does not have to be redeveloped every time an employee is to be trained or retrained. With a thorough JSA as a guide, all the steps of the process are identified and will not be overlooked during the training. The JSA also provides a basis for documentation of ongoing employee safety training. Training sessions with workers can be periodically conducted using the JSA to reinforce how to safely conduct a job.

A JSA can be of great value to the person investigating an accident. By reviewing the process and understanding the hazard, controls, job steps, and safe practices that have already been defined and implemented, accident investigators can gain valuable insight leading to a better accident investigation. The actions of the employee can be compared to the recommended job actions in the JSA to determine if the employee was performing his job properly. The investigator should also consider whether the JSA itself was correct and if any possible hazards are missing from the JSA, which might have led to the accident in question.

JSAs may also help supervisors learn about the various jobs their workers perform. If the supervisors have not actually performed a specific job themselves, the JSA can guide them through process of that job. Because the JSA identifies what is involved in each job, step-by-step, the supervisor can learn in detail about each of the individual jobs they supervise.

JSA DEVELOPMENT

All jobs in an organization have common elements. They all require people performing at their best level of productivity, producing goods and services in a safe and efficient manner. Helping do that is what the JSA is all about.

Often, employers, foremen, supervisors, and safety professionals conduct JSAs that are subsequently reviewed by workers who perform the job. At other times, workers may discover a task on the job site which does not have a written JSA and may conduct their own JSA on the job site before beginning the task. This worker produced JSA should be documented for review by the safety manager and the supervisor. JSAs should be developed for the following:

- jobs with the highest injury or illness rates;
- jobs recognized by employees as being high hazard with the potential to cause severe or disabling injuries or illnesses, even if there is no history of previous accidents;
- jobs in which simple human error could lead to a severe accident or injury;
- jobs that are new to your operation or have undergone changes in processes or procedures; and
- jobs complex enough to require written instructions.

Preparing for a JSA

- Review company accident history. Review with employees the worksite's history of accidents, losses that required repair or replacement, and any "near misses"—events in which an accident or loss did not occur but could have. These events are indicators that the existing hazard controls (if any) may not be adequate and deserve more examination. Involve company employees. Asking employees what they consider high hazard or high-risk job assignments and involving them in the initial hazard ranking of all jobs being considered for JSAs and the development and review of JSAs, creates buy-in and provides the whole-picture view. The people who do a job have a unique understanding of the realities of the job, and this knowledge is invaluable for identifying hazards as well as effective and workable control measures. Involving employees helps minimize oversights, ensures a quality analysis, and gets workers to buy in to the solutions because they will share ownership in their HSE program.
- Conduct a preliminary job review. Discuss with employees the hazards they know exist in their current work and surroundings. Brainstorm with them for ideas to eliminate or control those hazards. If any hazards exist that pose an immediate danger to the workers' life or health, take immediate action to protect the workers.
- Any problems that can be corrected easily should be corrected as soon as possible. Do not wait to complete a JSA before correcting serious hazards. This will demonstrate a commitment to safety and health and clear the way for identifying all hazards that present unacceptable risks and evaluating types of hazard controls.

Development of a JSA is not a one-time activity; it is a process. JSAs need to be developed properly with solid input from those familiar with the job in question; identified control measures must be agreed upon and implemented effectively; and affected workers and supervisors must be trained on the information contain in the JSA.

Additionally, JSAs should be seen as living documents. Keeping JSAs up-to-date may sound like a lot of extra work, but changes are not needed frequently. A company's safety program should have a system for regularly reviewing and updating JSAs. If a situation, a job environment, or any other change occurs, the change must be reflected on the JSA as quickly as possible. If the JSA is changed or revised, everyone concerned with the job should be informed of the changes and instructed in the new procedure.

When should supervisors review a JSA for possible updating? A JSA should be reviewed when an accident occurs on a job covered by the document. In this case, review the JSA to decide whether it needs revision. Check the JSA to see if the accident occurred because the JSA was not followed, or if the accident occurred because something was basically wrong with the job procedure or the analysis, or if a hazard existed that had not previously been recognized or eliminated.

When conditions change, such as job requirements, site conditions, manpower, or equipment operations, it is important to stop and reanalyze the job for potential new hazards created by these changes. New controlling measures should be put in place to eliminate or minimize the new hazard. If new controls cannot be implemented to reduce the hazard to an acceptable risk level, new engineering and administrative controls may need to be devised by job supervisors before work is resumed.

Writing the JSA: 3 Basic Steps

There are three basic steps in writing the JSA:

1. Each job is broken down into a sequence of steps. Each step describes the actions of the job as that job is performed.
2. Each step is examined to identify and define hazards (i.e., actions, conditions, possibilities) that could lead to an accident.
3. Recommended actions or procedures are determined for each hazard. The JSA becomes a guideline for what actions are necessary to eliminate or minimize the hazards that could lead to an accident or injury.

Step 1: List job tasks. Nearly every job can be broken down into job tasks or steps. Outline those sequential steps that it takes to carry out the job. Begin a job safety analysis by watching the employee perform the job and listing each step as it is performed. (Figure 2-1) Be sure to record enough information to describe each action without getting overly detailed. Avoid breaking down the steps in such detail that it becomes unnecessarily long or so broad that it does not include basic steps. It may be valuable to get input from other workers who have performed the same job. Later, review the job steps with the employee to ensure everything significant is included.

Point out that the analysis is done to understand the sequence of steps that it takes to conduct the job and is not an evaluation of the employee's job performance. Include the employee in all phases of the analysis—from reviewing the job steps and procedures to discussing uncontrolled hazards and recommended solutions. Be sure to document the findings in order to create a written record of the JSA.

Sometimes it may be helpful to photograph or videotape the worker performing the job or to photograph the work equipment and work environment. These visual records can be handy references when doing a more detailed analysis of the work. Management and workers may also find it useful to assign a probability and severity ranking to each hazard in the job, noting how likely or probable the hazard is to occur, and the severity of the consequences should it occur.

Step 2: Look for hazards. Once the basic steps of the job have been recorded, the next step is to identify the potential hazards associated with each step. Consider talking with employees who have performed the job for several years, as they usually have a wealth of information and will be able to describe injuries or near-misses they have had conducting the job, or injuries coworkers have experienced.

Source: Gorodenkoff/Shutterstock.com

Figure 2-1 Identify potential hazards

Do not just look for obvious hazards like a pallet leaning against the wall in an aisle. Look at the total environment. Some potential hazards are not obvious. Always consider the possibility of debris or other obstructions in aisles even if nothing is there at the time the JSA is developed. In order to consider the entire work environment, ask the following questions:

- What can go wrong?
- What are the consequences?
- How could the hazard arise?
- What are other contributing factors?
- How likely is it that the hazard will occur?

Document the answers to these questions in a consistent manner. Describing a hazard in this way helps ensure that efforts to eliminate the hazard and implement hazard controls target the most important contributors to the hazard. Probe for details regarding the job steps and the work involved in doing the job: How often does the job need to be done (e.g., daily, weekly, annually)? How often is each step of the job carried out (e.g., per shift, per hour)? How heavy is equipment that is lifted, pushed, or pulled?

Good hazard scenarios describe

- where it is happening (environment);
- to whom or what it is happening (exposure);
- what precipitates the hazard (trigger);
- the outcome that would occur should it happen (consequence); and
- any other contributing factors.

Table 2-1 Types of hazards to consider when conducting a JSA

Chemical	Biological	Ergonomic	Physical
Inhalation	Bloodborne pathogens	Awkward positions	Caught in/on/ between; pinch points
Skin contact	Building-related illness	Contact stress	Electrical
Absorption	Legionnaires' disease	Forceful exertions	Fire or explosion
Injection	Mold	Repetition	Noise
Ingestion	Plant and insect poisons	Vibration	Radiation
	Tuberculosis (TB)	Work area design	Thermal stress
	Wastewater		Slips and falls
			Striking against
			Struck by

The purpose of the JSA is to identify *all* hazards—both those produced by the environment or workplace conditions and those connected with the job procedure. Examine each step carefully from both perspectives to find and identify hazards (Table 2-1).

Practical questions to ask that will help identify potential hazards include the following:

- Can any body part get caught in or between two objects?
- Do tools, machines or equipment present any hazards?
- Can the worker make harmful contact with moving objects?
- Can the worker slip, trip, or fall?
- Can the worker suffer strain from lifting, pushing, or pulling?
- Is the worker exposed to extreme heat or cold?
- Is excessive noise or vibration a problem?
- Is there a danger from falling objects?
- Is excessive brightness or poor visibility a problem?
- Can weather conditions affect safety?
- Is harmful radiation a possibility?
- Can contact be made with hot, toxic, or caustic substances?
- Are there dust particles, fumes, mists, or vapors in the air?
- Can the worker make plant, insect, or animal contact?
- Can any foreign object contact the eyes?

A safety data sheet (SDS) can be good resource for assessing chemical hazards. An SDS is a written document that outlines information and procedures for handling and working with chemicals. SDS documents contain physical and chemical property information, potential hazard information, emergency procedures, personal protection and manufacturer contact information. For a knowledgeable user, exposure limits and chemical property data are the most useful because identifying a type of hazard alone does not inform you of severity. For example, hydrochloric acid and ferric chloride are both corrosive, but one can immediately damage on contact with skin or eyes and off gas to cause inhalation exposure and lung damage. Protective procedures also vary widely—some are generic or are more appropriate for uses with the highest potential for contact and exposure.

Step 3: Recommended controls and actions. Once the basic steps of the job and potential hazards associated with each step have been recorded, the final step in conducting the JSA is to determine ways to eliminate or control the identified hazards. A control measure is anything that will help to control the hazard by either preventing it from occurring or minimizing its impact if it does occur. If a hazard cannot be eliminated, steps should be taken so the consequences of the hazard are as low as reasonably practical in order to protect workers and the public.

When evaluating options for corrective measures, consider the hierarchy of controls:

- Elimination
- Substitution
- Engineered
- Administrative
- Personal protective equipment

As shown in Figure 2-2, progressing down the hierarchy results in reduced effectiveness. According to the Centers for Disease Control and Prevention, which developed the hierarchy, "The idea behind this hierarchy is that the control methods at the top of the list are potentially more effective and protective than those at the bottom. Following the hierarchy normally leads to the implementation of inherently safer systems, ones where the risk of illness or injury has been substantially reduced" (CDC 2010).

Elimination of the hazard altogether is the most effective control measure to reduce risk. By adopting the Planning through Design methodology described in chapter 4, many hazards can be eliminated or designed out of a system. For example, human interaction with potentially hazardous machinery or equipment can be reduced by automating certain processes. Similarly, the design team or operations and maintenance (O&M) staff can opt for an appropriate material or process substitution that is less hazardous. It is important to equally assess the substituted material or process for any risks it may pose.

Where elimination or substitution controls are not practical, the next strategy is to control the hazard at its source. Engineered controls do this, unlike controls that generally focus on the employee exposure to the hazard. The basic concept behind engineered controls is that, to the extent feasible, the work environment and the job itself are designed

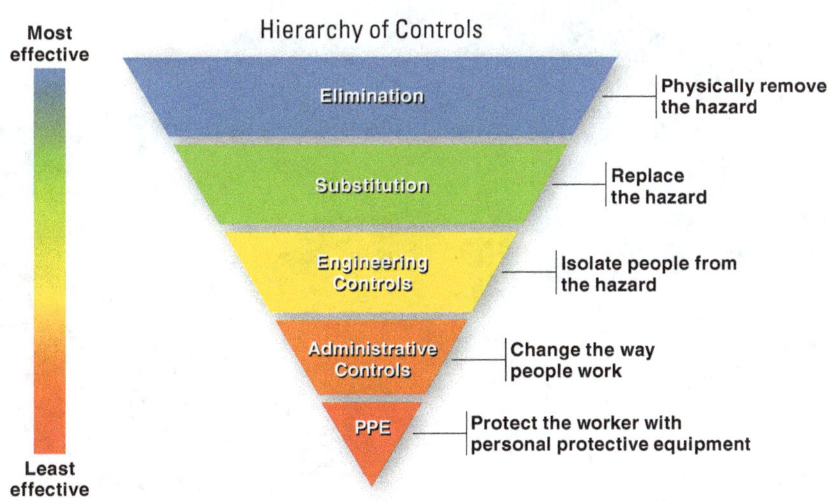

Source: Centers for Disease Control and Prevention (www.cdc.gov/niosh/topics/hierarchy/default.html)

Figure 2-2 Hierarchy of controls: apply the highest level of control commensurate with the risk level

to eliminate hazards or reduce exposure to hazards. Engineered controls can be simple in some cases. They are based on the following principles:

- If feasible, design the facility, equipment, or process to remove the hazard or substitute something that is not hazardous.
- If removal is not feasible, enclose the hazard to prevent exposure during normal operations.
- Where an enclosure is not feasible, establish barriers or local ventilation to reduce exposure to the hazard during normal operations.

If engineered controls are not feasible, administrative controls should be considered. Administrative controls include safe work practices, such as the utility's general workplace rules, standard operating procedures, or other operation-specific regulations. For example, even when a hazard is enclosed, exposure can occur during maintenance of the enclosed equipment. Through established safe work practices that incorporate additional protective measures when conducting maintenance operations, employee exposure to hazards can be further reduced. Administrative control measures also include additional relief worker, exercise breaks, and rotation of workers. These types of control are normally used in conjunction with other controls that more directly prevent or control exposure to the hazard.

When exposure to hazards cannot be engineered out of normal operations or maintenance work, and when safer work practices and other forms of administrative controls cannot provide sufficient additional protection, a supplementary method of control is the use of protective clothing or equipment. This is collectively called personal protective equipment (PPE).

While PPE (Figure 2-3) seems to be the most common form of protection and in some cases is mandated by law, it should be considered a last line of protection and always used

Figure 2-3 Personal protective equipment

in conjunction with engineered and administrative controls. PPE may also be appropriate for controlling hazards while engineered and work practice controls are being installed.

Note, one of the major weaknesses of PPE is the inconsistency of its effectiveness. This is primarily due to improper fit and application. Both of which can be hard to monitor on a routine basis.

As with Steps 1 and 2, document all the specific steps that need be done to correct or mitigate the identified hazards. Specify exactly what is to be done to correct the hazard, then take action to make those corrections. Using the JSA as a roadmap, follow through regularly to ensure that precautions are in place and new procedures are being followed. If new techniques for doing a job are required, provide training to affected employees.

The JSA can also be an effective means of uncovering health hazards in a job operation. The job could be contributing to a medical problem that otherwise might not become known for a long time. Examples are things like deterioration of the lower back, loss of hearing or loss of sight, chemical overexposure, or some other occupational disease.

These health problems are not visible or as dramatic as accidents because they usually do not happen all of a sudden. Nevertheless, they are extremely costly to the company and serious detriment to employee health. Many health hazards have to be assessed through measurements conducted by an industrial hygienist or other qualified individual. If a potential health hazard exists, it is important to determine the actual situation so appropriate steps can be taken to eliminate or control the hazard.

Challenges

Significant pitfalls are possible in developing JSAs that could result in a safety program being less effective than it might otherwise be. Common pitfalls include

- not listing all hazards;
- listing hazards but taking no action;
- listing nonspecific action and recommendations;
- failing to involve all levels of staff;
- failing to update JSAs when changes occur; and
- failing to train and retrain staff on JSAs.

THE JSA: AN IMPORTANT SAFETY FACTOR

If adopted effectively, JSAs can be valuable tools to help to reduce accidents and occupational illnesses and to increase operational efficiency. They often result in cost reductions, may lead staff to find a better way to do a job, and serve to document employee training. They are an important tool in the overall safety program.

Measurement of past incidents, successes and failures happens after the fact and is considered a "lagging" indicator. Measurement of future performance or commitment to tangible goals is considered a "leading indicator." Performing a JSA can help workers and management identifies potential hazards before they occur and implement corrections to minimize the risks. Setting tangible goals to perform safety analyses of all jobs, or to correct hazards so that they reach a specific minimal level of risk are examples of using leading indicator to drive a safety program, as opposed to lagging indicators, which measure past performance.

These items can be found in the electronic resources that are offered as part of this manual available at www.awwa.org/M3:

- Instructions for completing a Job Safety Analysis form

- Example of a completed Job Safety Analysis
- Job Safety Analysis—blank form
- Example of a job description and physical requirements for a conservation coordinator
- Example of a job description and physical requirements for waterworks operator field crew
- Hazard action list—blank form

REFERENCES

Centers for Disease Control and Prevention (CDC). 2010. The National Institute for Occupational Safety and Health (NIOSH) Hierarchy of Controls. www.cdc.gov/niosh/topics/hierarchy/default.html. Accessed December 15, 2021.

Michigan Department of Licensing and Regulatory Affairs, Michigan Occupational Safety & Health Administration, Consultation Education & Training (CET) Division, National Safety Council. 2020. Job Safety Analysis. SP #32. [Online.] Accessed December 15, 2021.

OSHA (Occupational Safety and Health Administration). 2002 (revised). Job Hazard Analysis OSHA 3071. [Online.] Accessed December 15, 2021.

Texas Department of Insurance, Division of Workers' Compensation Workplace Safety. Job Safety Analysis. [Online.] Accessed December 15, 2021.

Chapter 3

Incident Management

Mike Jacobs, Brian Exendine

Even after the most thorough Job Safety Analysis (JSA), incidents can still happen. The repercussions of any incident can affect a utility's financial stability. According to the Occupational Safety and Health Administration (OSHA), US employers pay an estimated $1 billion per week in direct workers' compensation costs alone. This does not include the sometimes-immeasurable indirect costs of incidents. These indirect costs include lost productivity, training replacement employees, incident investigation, and low employee morale, to name a few. As shown in Figure 3-1, OSHA has historically used the results of a 1981 Stanford University study that found the indirect costs of incidents on the job can range from 1.1 (for the most severe injuries) to 4.5 (for the least severe injuries) times the direct costs, which include medical expenses and insurance premiums for workers' compensation benefits, liability, and property losses. According to OSHA, "the lower the direct costs of an incident, the higher the ratio of indirect to direct costs" (US Department of Labor 2013).

INCIDENT MANAGEMENT PROGRAMS

In addition to JSAs, formal incident management programs are invaluable components of any comprehensive health, safety, and environmental (HSE) program. Incident management serves to reduce costs of future incidents by analyzing and improving upon utility processes. A Water Research Foundation (WRF) project, *Water Utility Safety and Health: Review of Best Practices* (Borowski and Adams 2010), found that "one of the most significant opportunities [in the water industry] for improving safety and health performance is to conduct thorough incident investigations and to manage the return to

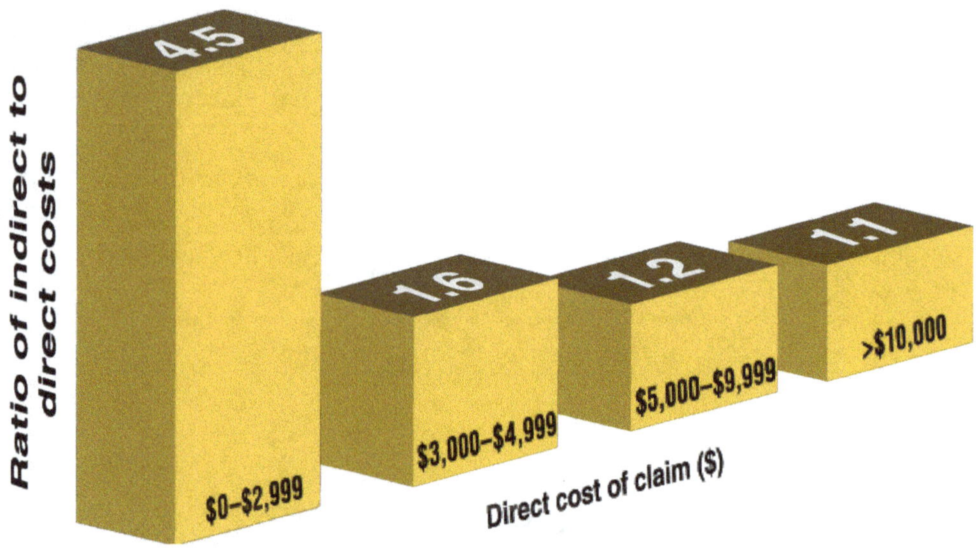

Figure 3-1 Ratio of indirect to direct costs of on-the-job incidents

work process for the injured workers." From initial emergency response and reporting to effective return- to-work (RTW) programs and record keeping, this chapter will outline practices that will help develop an efficient injury and illness management process.

Incident Response

Initial medical attention. A utility must be prepared to respond quickly and efficiently to any injury. The first priority at the scene of an incident is to ensure that the area is safe for the injured employee and the responders, and that the injured employee receives adequate medical attention. Employees should be assisted with obtaining, or instructed to seek or provide, medical attention as needed before doing anything else.

To facilitate a quick response, a utility should have readily available first aid kits and facilities that are appropriate to the employees' potential exposure to injury in any given area of the facility (e.g., eye wash in the laboratory). Employees who may need to provide immediate care (e.g., distribution or plant operators) should be trained in first aid/cardiopulmonary resuscitation (CPR) and safe handling of bloodborne pathogens. In some cases, this training is required (e.g., a confined-space attendant) and a utility should check local regulations. A review of completed Job Safety Analyses can help determine which employees should be trained in first aid/CPR; the number of employees per shift or per work area that are to be trained should also be outlined in the HSE program. All employees should be trained in how to contact emergency medical services. If emergency services cannot respond in a reasonable amount of time, a utility may need to establish an in-house emergency response team with specialized training in immediate medical care.

For injuries that do not need immediate care but do require medical attention, utilities should consider contracting with an occupational health provider (OHP). This will ensure that employees are seen in a timely manner by an occupational health doctor who is specially trained to deal with occupational injuries and illnesses. By establishing a relationship with an OHP, a utility also ensures that the treating physicians are familiar with the types of work being performed at the utility, as well as the utility's workers' compensation insurer's requirements and return-to-work program.

Incident Reporting

All incidents should be reported as soon as possible so each incident can be investigated thoroughly—before evidence is altered or destroyed—and causes corrected. Initial reports also provide documentation for incidents that were initially thought to require on-the-scene first aid only, but later require medical attention and reporting under workers' compensation requirements and/or must be logged for OSHA or another regulatory agency. Once an injury or illness is reported, the response will help determine the ongoing well- being of the employee and the ultimate cost of the claim.

Supervisors are typically responsible for investigating incidents and should be trained in post-incident response, even if the utility rarely has work-related injuries. An infrequent claim history makes it more likely that employees and supervisors will not know what to do in the event of an injury. They may inadvertently make a decision that could adversely impact the well-being of the employee and the length or cost of the claim. It is the responsibility of utility management to establish an incident-response program that clearly defines the roles and responsibilities of employees, supervisors, and any other staff that may be involved in the reporting and investigation process. A strong incident reporting and response program identifies, probes, and documents all aspects of an incident scene and the events that led to the incident.

After the appropriate medical attention is provided, incidents should be reported to the employee's immediate supervisor. The supervisor should then complete an injury report form that documents the initial findings. The form used for initially reporting an injury should be simple but informative. The form should include the who, what, when, and where, and how (i.e., it is not only important to know if personal protective equipment was being used, but also *how* was it being used) of the incident. A complete injury report will include the answers to the following questions:

1. What exact job was the injured employee doing? (e.g., Employee was loading water pump onto truck.)
2. What tools, materials, or equipment were being used to accomplish the work assignment?
3. What specific action caused the incident? (e.g., Employee slipped, and water pump hit his side.)
4. What action did the employee do or not do that caused the injury? (e.g., Employee was not using hand crane to load the pump.)
5. What safety protection devices were provided? Were they being used?
6. What, if any, unsafe conditions or defects in equipment, materials, or tools were there at the job site where the incident occurred?
7. What, if anything, was wrong with work methods employed or other acts that caused the incident? (e.g., The hand crane is provided to load and unload heavy items from the truck. Employee was in a hurry and did not use the crane.)
8. What safeguards were being used (or not) and what safeguards should have been used?
9. What steps will be taken to prevent similar injuries? (e.g., Incident was discussed with crew at weekly safety tailgate meeting.)
10. What else should be done to prevent recurrence? (e.g., Foreman or helper should check loading and unloading of equipment to ensure that safe procedures are followed.)
11. Were there any witnesses to the incident or the events leading up to it?

Some utilities may have other staff, such as a safety, risk, or human resource specialist, who can assist with the gathering of information and completion of the incident report. If the incident results in a serious injury that requires medical attention, a report will typically need to be filed with the utility's workers' compensation insurer. Regulatory agencies (e.g., OSHA) also have reporting requirements.

Investigation and Documentation

A thorough investigation following an incident provides an excellent opportunity for the utility to demonstrate that it values its employees' well-being and is interested in preventing future incidents. Findings from the investigation should be used to reinforce good safety practices, develop control methods, eliminate hazards, and develop training. A properly implemented incident investigation process will promote continuous improvement and can reduce the recurrence of injuries and cost of claims.

As stated above, the investigation should begin immediately after an incident occurs. The injured employee's supervisor is typically the most familiar with the work being done and the work area; the supervisor should initiate the investigation and ensure that important evidence is not moved, lost, taken, or destroyed. The site investigation should establish how the incident was work-related and document information that cannot be preserved but will be needed to complete a formal incident report. The investigation should also include photographing the work area and equipment involved as well as interviewing witnesses.

If the injury is minor (i.e., first aid only) and the cause is clear, the supervisor may be able to complete the investigation and report the findings quickly, with no need to contact outside authorities. If the injury is serious (i.e., requires medical attention) and/or the cause is unclear, a more thorough investigation will be needed. This investigation may involve employees, safety committee members, management staff, engineers, safety professionals, and insurance representatives. The site investigation should establish surface causes of the incident. Guidelines for conducting this investigation include the following:

- Record the time of day, location, and type of work being done at the scene.
- Use all of your senses and take note of your initial observations:
  - What is the condition of the equipment, tools, and other materials involved in the incident?
  - Are there skid marks, spills, stains?
  - Is there anything that seems irregular or out of place?
  - What are the weather conditions?
  - Is there excessive noise, light, or other distractions?
  - Who is on scene?
- Take photos. Some important points to remember:
  - Take photos at different angles.
  - Be sure to capture the entire scene and all equipment and witnesses.
  - Take notes on each photo. Identify the type of photo, date, time, location, subject, weather conditions, measurements, etc.
  - Place an item of known dimensions in the photo if hard-to-measure subjects are being photographed.
  - Identify the person taking the photos.
- If possible, videotape the scene. Video recordings can provide detail that photos may not be able to capture.

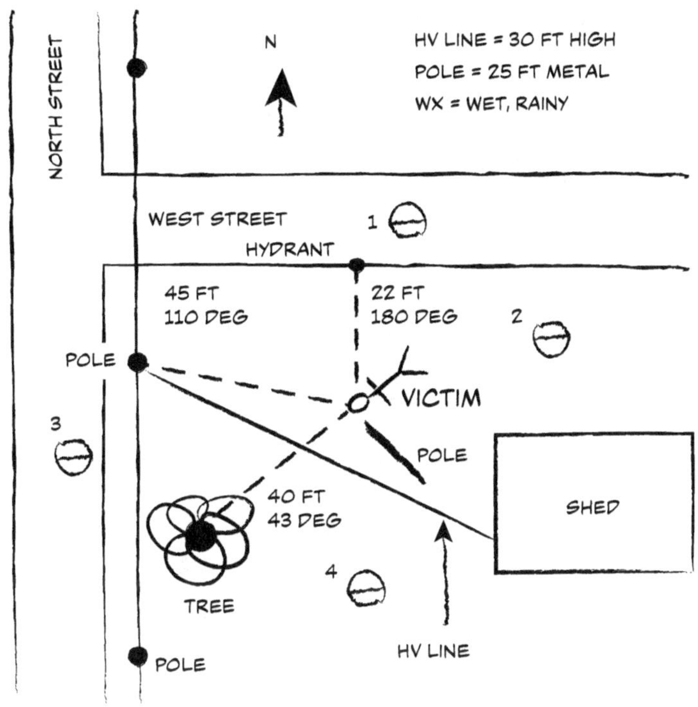

Source: Adapted from OSHA

Figure 3-2 Triangulation method

- Take statements from witnesses and ask them to explain how the incident happened. Also, have them describe anything that was moved or otherwise disturbed before the investigation began.

Measure distances and positions of anything and everything you believe to be of any value to the investigation and make a diagram of the incident scene. Be precise when making diagram and make sure measurements are tied to permanent points. Other information that should be included on the diagram includes direction (N, S, E, W) and where people were located when the incident occurred.

The diagram in Figure 3-2 illustrates the triangulation method, which makes it possible to later pinpoint the exact location of an object. In this incident, the victim contacted a high-voltage line with a metal tree-trimming pole. The position of the victim's head is measured from three points. Notice the small circles with horizontal lines through them. These circles indicate where photos were taken. Also, directional north is indicated, and all major objects are identified.

Figure 3-3 is another diagram that demonstrates a major advantage of illustrating an incident scene. It shows motion through time by diagraming how a bulldozer rolled down the side of a hill.

Root Cause Analysis

The initial and site investigations identify surface causes of the incident such as technical defects or employee error, but the investigation must not stop there. If it does, the opportunity to prevent similar incidents may be lost. Oregon OSHA fatality investigator Mike Riffe says it best: "Too often, finger pointing is used in incident investigations. If the employer's goal is to find out who did it, they will never get to prevention. They need to

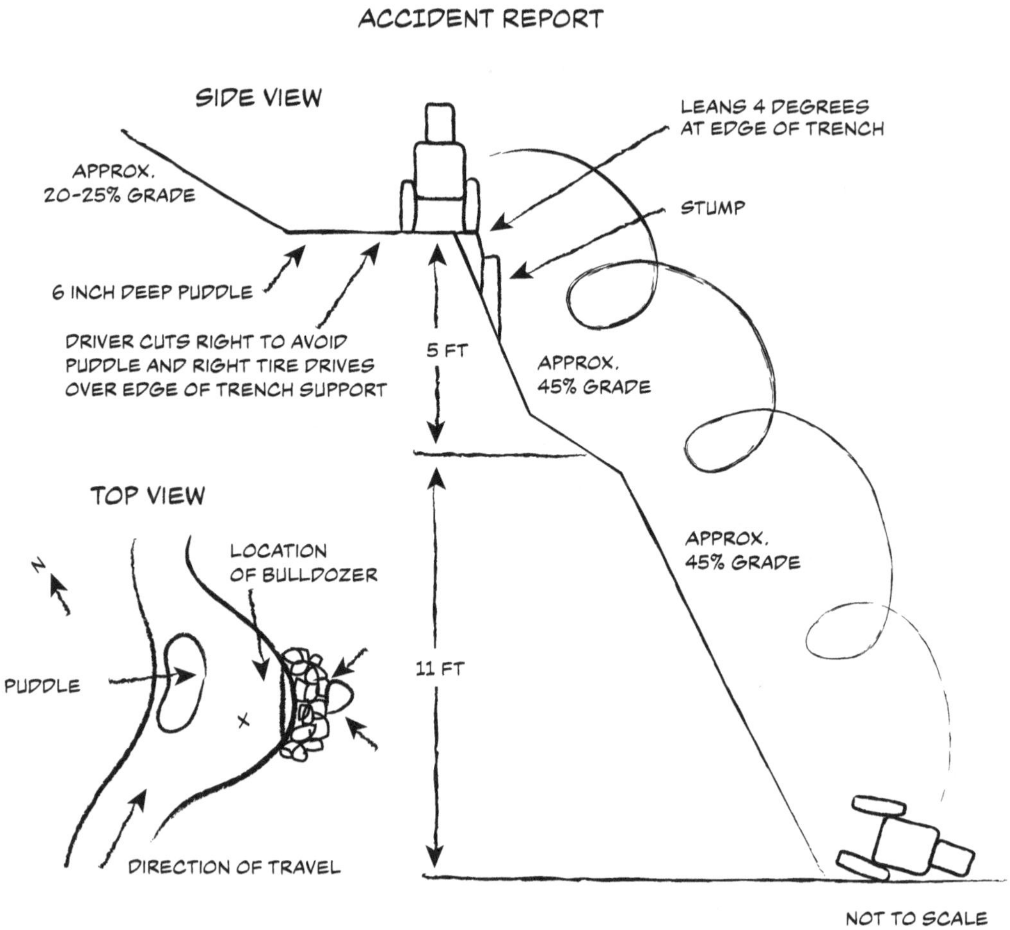

Source: Adapted from OSHA

Figure 3-3 Sketching motion through time

ask, 'What went wrong?' and then look carefully for weaknesses or holes in their safety program and fix them to prevent future incidents" (Mesaros 2012).

When technical defects and employee error contribute to an incident, they are often linked to underlying safety system conditions that have created a hazardous situation and unsafe behaviors that are manifested in the surface causes of incidents. These underlying conditions are typically the root cause of the incident. The "incident weed" shown in Figure 3-4 is often used to illustrate the relationship between surface and root causes, and common examples of each are shown.

Many techniques are used to evaluate the root causes of incidents, but one of the most useful is the "5-Why" technique, which is relatively easy to use and requires no advanced statistical analysis or data collection. The technique follows a structured question-and-answer sequence that contains a feedback loop to the previous question. By asking enough times why an event happened, the technique is supposed to "peel away the layers of symptoms" associated with a problem. The process continues until another answer cannot be justified and the root cause is narrowed to the last question or answer set. An example of a question-and-answer sequence using this technique is as follows:

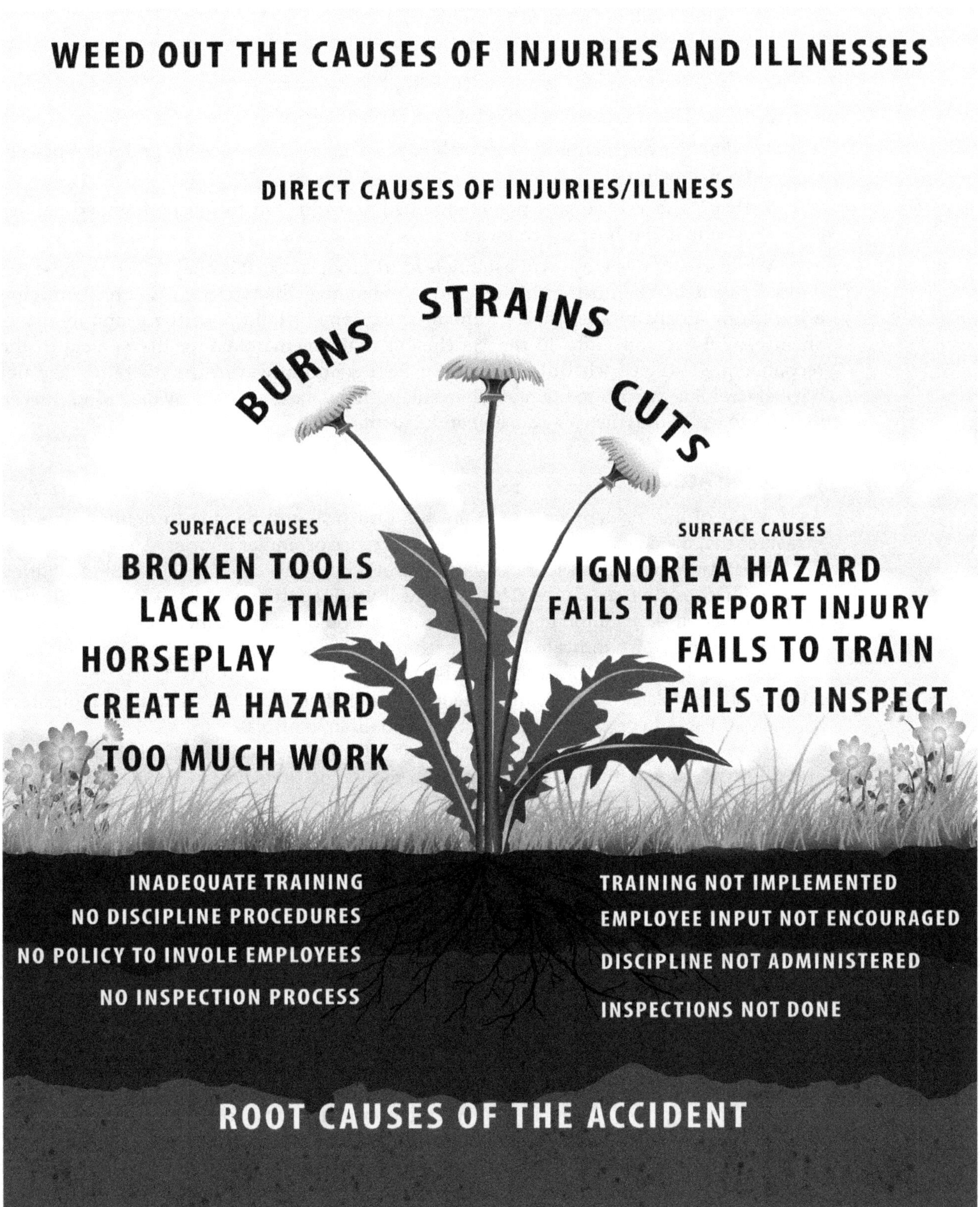

Source: Oregon OSHA Instructors Manual: Accident and Incident Investigation

Figure 3-4 Incident weed and root cause analysis

The vehicle will not start. (The problem)

1. *Why?* The battery is dead. (First why)
2. *Why?* The alternator is not functioning. (Second why)
3. *Why?* The alternator belt has broken. (Third why)
4. *Why?* The alternator belt was well beyond its useful service life and not replaced. (Fourth why)
5. *Why?* The vehicle was not maintained according to the recommended service schedule. (Fifth why, a root cause)

While asking five whys will usually lead to a root cause, it is okay to ask six or seven whys if needed. When using this technique, remember that the focus is on identifying a broken or absent process, not a human or equipment failure. In the example above, an investigator could jump to the conclusion that not maintaining the vehicle to the recommended service schedule is operator negligence; however, the operator may not have the training or parts to maintain the vehicle. In this case, a sixth why may be necessary in order to get to the true root cause of an incident.

Corrective Action

As a result of the incident investigation, the utility should be able to define corrective measures that need be taken to prevent the recurrence of similar injuries. As in developing the Job Safety Analyses, the hierarchy of controls should be considered when evaluating options for corrective measures. The incident investigation may alert O&M or HSE staff to risks that could be eliminated or substituted through site modifications. The incident investigation may also indicate that an employee was not wearing appropriate personal protective equipment (PPE). Although use of PPE should be enforced, especially while additional controls are being implemented, remember that PPE is only a temporary solution and should be used in conjunction with other controls.

When recommending corrective actions, consider the decision maker's perspective and come prepared with the appropriate information. In preparing recommendations, the following information should be provided:

- What is the hazard?
- What is the incident and incident history associated with the hazard?
- What is the recommended corrective action?

Be prepared to provide a simple cost–benefit analysis. To provide this information, research the direct cost (discussed in the Introduction to this chapter) associated with this incident and/or similar incidents, and the potential indirect costs. Compare these costs to the projected cost of the corrective action.

Once corrective action has been approved, the responsibility to take those actions by an assigned date should be given to a specific employee, and the actions' implementation, activation, and completion should be tracked.

The finding from the incident investigation along with corrective measures to be taken should be shared with all employees and management (note that before sharing this information the name of the injured employee should be removed). Typically, this information can be shared through the safety committee.

A final future step is to evaluate whether the corrective action has been effective at reducing that type of incident, or effective in promoting employee awareness or learning, or other actions.

Return-to-Work Programs

A return-to-work (RTW) program is a formal process to return injured employees to work with restrictions while they continue their recovery. According to the WRF project *Water Utility Safety and Health: Review of Best Practices* (Borowski and Adams 2010), "Water utilities often offer employee benefits that include liberal leave policies. These policies compromise employers' ability to effectively manage injury cases." The study also found that many utilities are unwilling to offer light-duty and RTW programs. This may be because utility management is not fully aware of the benefits that a RTW program can offer.

In fact, studies have shown that injured workers who are off work longer than six months have only a 50 percent chance of returning to their jobs. If time lost exceeds one year, chances are 90 percent that they will never return to work (Chandler 2003). Returning an employee to light duty is much cheaper than losing an employee completely to an injury.

Workers' compensation carriers can be a great resource when developing a RTW program. Carriers can provide sample programs and program review. Also, some states or workers' compensation carriers may provide incentives for returning an employee to work; for example, Oregon will pay 50 percent of a worker's wages while that employee is on light duty (up to 66 days) and pay for work-site modifications that may be needed so the employee can go back to work safely.

The following are key aspects of an effective return-to-work program.

Management support. The only way a return-to-work program can work is if it is supported by upper management. A utility should have a written RTW program that is agreed upon by its entire management team. The program should demonstrate the utility's commitment to returning employees to work after an injury, as well as define both the utility's and the employee's responsibilities. This will ensure that, regardless of varying circumstances, the process will be followed.

Knowledge of physical requirements. In order to determine whether an employee will be able to return to work after an incident, the physical requirements of that employee's job must be understood. As with a JSA, each position should be evaluated and the essential functions of the job, including the job's physical requirements, should be documented. Most positions can be evaluated by the immediate supervisor or manager. However, positions with extreme physical requirements may need to be evaluated by a safety professional or physical therapist.

For positions with extreme physical requirements, the JSA should be used to develop a physical capacity test (PCT) that measures an employee's ability to perform simulated or real job activities. Employees should be required to pass a PCT along with a pre-placement physical to ensure that they are capable of performing the physical requirements of the job. The PCT can also be used to evaluate an employee's ability to return to work.

Identification of light duty. After job descriptions have been developed, duties within the job should be identified that are physically less strenuous. This will help when an employee with restrictions is trying to return to work. Although it is usually more beneficial to return employees to light duty within their positions, sometimes their restrictions do not allow it. It may be necessary to have them work with another department.

Occupational health provider. The treating physician is, of course, primarily responsible for applying the most appropriate clinical care to treat the injury or illness. However, the physician is often the key to minimizing the time away from work as well. Depending on local laws, employees may be able to see a physician of their choice. Regardless, a medical facility that understands occupational medicine should be available. Select a nearby occupational health facility, so injured employees can receive quick attention. For traumatic or after-hours injuries, the local emergency room may be more appropriate.

Using the same medical facility to conduct routine services, such as pre-placement physicals and physical capacity testing, can be beneficial. For these routine services, medical staff with a good picture of utility operations will be more in tune with the physical requirements expected of employees. It may be appropriate to invite staff from the facility to see some of the more specialized tasks.

Claims management. Returning an employee to work should be the responsibility of one person, a claims manager, within the utility. While many will be involved in the process, possibly including a professional claims manager or third-party administrator, having one person in the utility responsible will ensure that the process is effective and consistent.

The claims manager should have a thorough understanding of the RTW process and any applicable laws. To maintain an effective RTW program the claims manager should

- Maintain the written RTW program and ensure that everyone has the appropriate training.
- Communicate with the injured employee on a regular basis, ensuring that he or she understands and is comfortable with the process.
- Interact with the treating physician and provide the doctor with details of the utility's RTW program, including available light duty. Also ensure the physician receives the injured employee's on-the-job physical requirements.
- Ensure that the physician provides detailed restrictions and that those restrictions are followed when the injured employee returns to work.
- Keep the employee's supervisor and manager informed of the injured employee's status.
- Interact with the utility's insurance company to ensure all of the required information is provided in a timely manner.
- Maintain a file of all claims-related records.

Record Keeping

Maintaining records of work-related injuries and illnesses is not only a good idea, it is required by OSHA and state and provincial agencies. In the case of OSHA, records of injuries and illness that are work-related must be kept on OSHA's Form 300, Log of Work-Related Injuries and Illnesses. OSHA inspectors will typically request these records when performing an inspection.

In most cases only records of significant injuries and illness (those requiring medical attention from a physician) are required to be kept. However, maintaining records of less significant injuries and even near misses can be beneficial. These records can be used in a proactive way by the utility to

- manage company safety and health programs;
- locate workplace hazards;
- raise employee awareness regarding safety and health issues;
- establish the need for additional employee training; and
- identify injury and illness trends.

The following electronic resources are available at www.awwa.org/M3:

- Incident Investigation Program Example
- Incident/Incident Analysis – Blank Form
- Early Return to Work Program Example
- Physician Release-to-Work – Blank Form

- Temporary Modified Duty Job Description Example and Form
- Example of a job description and physical requirements for a conservation coordinator
- Preplacement Physical Capacities Test for Construction Crew – Blank Form
- Example of a job description and physical requirements for waterworks operator field crew
- Passing Criteria for Maintenance, Valve, Construction Crew Fitness Example
- Physical Capacity Requirements – Blank Form

REFERENCES AND BIBLIOGRAPHY

Borowski, J.H. and Adams, P. 2010. *Water Utility Safety and Health: Review of Best Practices.* Water Research Foundation Project No. 3104. Denver, Colo.: Water Research Foundation.

Bose, H.A. 2008. Returning Injured Employees to Work: A Review of Current Strategies and Concerns. *Professional Safety.* June 2008, 63–68. [Online.] Accessed December 15, 2021.

Chandler, D.L. 2003. Constructing Return to Work Programs: Building for Better Returns, A Public Entity Risk Institute (PERI) Symposium.

Mesaros, M. 2012. "Making Sense of Tragedy: An Oregon OSHA Investigator's Mission." *Oregon Health and Safety Resource.* [Online.] Accessed December 15, 2021.

Oregon OSHA. Accident and Incident Investigation Instructor Guide. [Online.] Accessed December 15, 2021.

Oregon OSHA Online Course 1110. Conducting an Accident Investigation. [Online.] Accessed December 15, 2021.

This page intentionally blank.

Chapter 4

Prevention Through Design

Mia Parton

According to the AWWA Water Sector Utility Health, Safety and Environmental (HSE) Program Survey completed in 2018, 62 percent of utilities involve HSE operations and maintenance staff in the design phase of new or renovated facilities to help reduce risks in the workplace by designing out hazards and designing in safety controls (i.e., Prevention through Design [PtD]). In addition, 53 percent of utilities use internal design standards, and 38 percent of utilities use proactive design safety reviews as methods and tools to minimize hazards and risks.

Investigations conducted through incident management programs often lead to changes in procedures, training or easy to implement process/facility changes, creating an incrementally safer work environment. Construction of a new or extensively renovated facility also provides the opportunity for a fundamental reconsideration of utility processes, and facilities to eliminate hazards and create inherently safer facilities. Both new construction and extensive renovations benefit from addressing safety factors early on in the design phase, both to minimize cost and allow more fundamental improvements. This chapter outlines how HSE; engineering; and operations and maintenance (O&M) teams can collaborate to effectively reduce risk in the workplace by designing out hazards and designing in safety controls.

WHAT IS PREVENTION THROUGH DESIGN?

Per the National Institute for Occupational Safety and Health (NIOSH), one of the best practices to prevent and control occupational injuries, illnesses, and fatalities is to "design out" or minimize hazards and risks. NIOSH leads a national initiative called Prevention through Design (PtD). NIOSH's PtD initiative was preceded by similar voluntary efforts in other organizations (e.g., numerous AIChE Process Safety Management guidelines and ANSI/ASSP Z590.3) and formal regulations in some countries (e.g., OSHA standard for highly hazardous chemicals and more broadly in the United Kingdom). The NIOSH mission for the PtD initiative is "to prevent or reduce occupational injuries, illnesses, and fatalities through the inclusion of prevention considerations in all designs that impact workers. The mission can be achieved by

- eliminating hazards and controlling risks to workers to an acceptable level 'at the source' or as early as possible in the life cycle of items or workplaces;
- including design, redesign, and retrofit of new and existing work premises, structures, tools, facilities, equipment, machinery, products, substances, work processes, and the organization of work; and
- enhancing the work environment through the inclusion of prevention methods in all designs that impact workers and others on the premises" (NIOSH 2013).

PtD is the systematic assessment and mitigation of facility hazards that can arise during construction, operation, and maintenance. By identifying hazards early in the project, facility, or process design and by incorporating design changes and controls, the projects can be built without those hazards, or with hazard reduction measures included. PtD is a proactive process that brings together knowledgeable health and safety, engineering, operations, and maintenance staff in the conceptual and early design phases. Through a risk-identification process and subsequent application of the hierarchy of controls, water utility facilities can be designed from the ground up with fewer hazards and with control measures already included.

The objective is to incorporate safety solutions as early as possible during initial planning and design, when fundamental choices can best be made without disrupting schedule or incurring unnecessary cost, to maximize safety performance during construction, operations, and maintenance activities. There are two parts to PtD: reduce the risk of hazards before they happen and mitigate their impacts if they do happen. In the long term, it can help utilities operate their facilities more cost-effectively if it can. PtD thereby reduces the risk of injuries and the need for complicated, expensive procedures and/or personal protective equipment (PPE). Examples of hazard elimination in early scoping and conceptual design stages is alternative chemistry or inherently safer chemicals. Later reviews may identify additional hazard reduction options, and as such, PtD can be an effective methodology for not only new design, but for renovations, additions, and site modifications as well. For example, if a potential permit-required confined space can be modified through improved access, lighting, ventilation, and/or other hazard-reduction methods, work in that space may no longer be classified as a permit-required confined-space entry. Monitoring, rescue, ventilation, PPE, and so on will only be required based on the job or task hazards. The safety features included at the design stage will benefit the workers for years to come when that facility goes on-line. The PtD concept is to look beyond designing just for minimum regulatory compliance. Instead, all safety hazards that can be designed *out*, or practicable safety controls that can be designed *in*, are identified and included by the design team. For example:

- A vertical ladder inside an elevated tank with a rail fall protection system can be more difficult to use for maintenance workers if is angled (it would require additional strength), or if there is a big gap from the ladder to the catwalk (could make the transition more difficult and therefore become more hazardous).
- In larger valve vaults with a wheel to open and close the valve, if the maintenance worker has to climb up the pipe to operate the valve vs. having the valve wheel

installed to the side for easy access, may increase chances of worker getting injured by falling or slipping.

- A potential permit-required confined space can be eliminated (e.g., well house s. vault) or modified through improved access, lighting, ventilation, and/or other hazard reduction methods, so work in the space may no longer be classified as permit-required confined-space entry. Monitoring, rescue, ventilation, PPE, and so on, will only be required if the job or task introduces new hazards into space.
- If permits and security is available, especially inside a locked facility, consider avoiding underground confined spaces, by raising the pipes and valves above ground for easy access to both reduce the hazards and the number of maintenance employees to perform the work safely and effectively.
- Avoid roof fall protection requirements by designing buildings with a flat roof and parapet meeting the railing requirements.

In general, installing remotely controlled processes that would reduce the need for manual operations that also reduces the risk and hazards.

PTD IMPLEMENTATION IN THE WATER INDUSTRY

Water transmission, treatment, and distribution facilities and systems present many opportunities for PtD, such as

- moving a valve actuator above-grade, out of a valve vault, to eliminate the need for operators to enter a permit-required confined space to operate the valve;
- providing a platform with appropriate access and railings so that workers are at the proper ergonomic height while operating a valve or performing maintenance and repair, and workers do not need to work from a ladder;
- providing adequate lighting at facilities to increase visibility for workers;
- installing monitors, sensors, and alarms to notify workers of any hazardous environments in advance prior to entering hazardous facilities;
- considering smart city technologies such as artificial intelligence (AI), not only is it possible to automate to reduce operational risk, but also to simulate the behavior of water networks under different scenarios and detect anomalies, thus improving efficiency;
- when possible, utilize remotely controlled under water cameras for condition assessments and inspections of tunnels, water tanks and or basins;
- utilizing virtual and augmented reality technologies for training of employees, which enables optimizing the operation and maintenance of the facility and creating more effective learning environments for operators;
- using Building Information Modeling (BIM) technology to enable owners to leverage facility data through the facility lifecycle to provide safe, healthy, effective, and efficient work environments;
- outsourcing of work with high risk to external sources which have the trained specialist, technologies, and resources to deal with such hazards;
- specifying the purchase of quieter equipment so that employees can work in the area without the requirement for hearing protection use; and
- providing equipment to lift and feed treatment chemicals automatically, eliminating the need for workers to handle heavy bags of chemicals. PtD and maintaining

safety over the life of the facility can be fostered through the organization's project planning and implementation processes as well, for example:

- requiring record drawings documentation upon completion of construction is essential for safer maintenance, rehabilitation, and reconstruction in the future;
- inspection and auditing construction contractors, as well as requiring Job Safety Analysis (JSA) for hazardous tasks, can reduce incidents during construction; and
- requiring ongoing feedback from the operations and maintenance and inspection staff during design to identify and incorporate opportunities for improvements.

Roles and Responsibilities

Effective application of PtD requires a team approach, including representation from HSE, engineering, and O&M professionals. The best design incorporates input from all team members providing their different perspectives, technical knowledge, standards and criteria knowledge, and knowledge of actual work practices in the field. Field knowledge from employees who work or have worked hands-on with the equipment or processes is essential. Cost estimators may also be part of the PtD team to provide financial analyses of options, returns on investment, and other costs. This approach has been practiced in many industries and other areas as a central mechanism for simulating innovation. All involved parties are encouraged to provide their feedback and input on areas of improvement in the design or processes. This can be done by requiring specific division such as O&M and inspections to participate in the review process of the design. This can be under the HSE management program policy or Quality Assurance and Quality Control (QA/QC) plan. Some of the advantages of OFI methods are as follows:

- There is a mechanism for employees to be more involved in HSE and projects designs.
- There is a systematic way to understand the safety problems as each project is unique and can bring new challenges.
- More management involvement which is the center of quality improvement and is the motivator for innovation and thinking outside of the box to create more efficient process and improve workplace safety.

When identifying PtD roles and responsibilities, consider how and why a particular safety concern is under discussion. For design of a new project or process, the design team, led by engineering, will likely perform alternatives analyses and prepare conceptual engineering reports, and consequently will need to initiate PtD review for design alternatives. However, if a safety concern is identified by HSE professionals as part of an injury analysis, that team will likely lead the initial PtD effort in the event that safety concerns necessitate major renovations or site modifications.

Typical responsibilities for each sector include:

Health, safety and environmental

- Identify hazards of existing facility designs or processes that could be reduced or eliminated through the PtD process.
- Conduct hazard analyses (using various methodologies such as Job Safety Analysis) to identify risks and prioritize them.
- Identify hazard controls and/or reduction strategies.

- Participate in reviews to identify and rank PtD design alternatives.
- Conduct design review of projects to identify hazards, controls, and opportunities to implement risk-reduction measures, and communicate information to design team.
- Coordinate with O&M staff to reach consensus on best PtD measures.
- Participate on post-project lessons-learned reviews for effectiveness of PtD measures.

Engineering and project management

- Communicate and coordinate with other PtD stakeholders (construction manager, security, environmental, purchasing, public relations, finance, IT) early in any new project or process design, or during upgrade of existing operations.
- Consider alternative chemicals, processes, sites, and design approaches at the earliest conceptual stage to determine if inherent safety choices can be cost effectively made.
- Provide HSE and O&M with sufficient technical information on process/project alternatives to enable hazard evaluation.
- Implement standard PtD design criteria as determined by the agency.
- When PtD design challenges arise, coordinate discussion with PtD team members to resolve issues.
- Include PtD considerations in conducting alternatives analyses and ranking alternatives.
- Provide outside designers with PtD criteria when projects are not designed by in-house teams.
- Provide regular design review opportunities for stakeholders, including O&M and HSE, and incorporate PtD feedback.
- Hold lessons-learned sessions at project closeout to gather information on the success of PtD solutions or the need to make changes in the future.

Operations and maintenance

- Identify hazards of existing facilities and processes, including ideas for improvements on an ongoing basis.
- Participate in formal and informal hazard analysis processes, providing key input from the user perspective.
- Participate in review of design alternatives that incorporate PtD measures, providing input from the user perspective.
- Participate in the design review process, identifying health and safety hazards and PtD opportunities to the design team.
- Participate in post-project lessons-learned reviews of PtD measures' effectiveness.

For example, inherent safety for the installation of a new tank would include evaluation of alternative sites where a ground tank may be viable instead of an elevated tank and brainstorming of design options for a ground tank that might eliminate the need to routinely climb to the top and provide appropriate safeguards for when climbing will be necessary. A design of rehabilitation of an elevated tank, would include an HSE plan that would include at least one representative from Engineering and Project Management, O&M, and HSE to review the construction drawings during the conceptual, preliminary and final stages of design, utilize innovative and effective solutions to not only improve the safety of worker during construction and maintenance but should also integrate value engineering for reducing the capital and ongoing costs of the operations of the facility.

PtD Methodology

Hierarchy of controls. When implementing PtD, the hierarchy of controls provides a systematic structure to rank the preferred safety approach(es) and determine the best approach, given the project-specific parameters. Consider the six levels of the hierarchy of controls described in Chapter 2, Figure 2-2:

1. Elimination
2. Substitution
3. Engineering controls
4. Administrative controls
5. Behavior
6. Personal protective equipment (PPE)

The PtD focus is to achieve workplace safety by levels 1–3, thereby avoiding reliance on levels 4–6. While the optimal approach is complete elimination of a hazard, there may be limitations because water industry facilities have some inherent hazards that cannot be completely designed out. Therefore, the PtD goal is to use the hierarchy of controls to modify the design within practicable limits, reducing the hazards as low as is reasonably possible.

Hazard identification and evaluation methods and tools. The process of identifying hazards/risks can range from informal, empirical methods to extensive, formal hazard evaluations such as a JSA. Water industry staff should consider the scale and scope of the project under consideration when selecting the appropriate hazard evaluation method. Specific hazard evaluation methods are covered in detail in ANSI/ASSP Z590.3 – 2011. In general, methods may include

- JSA;
- design checklists;
- process hazard analysis (e.g., what-if analysis, checklist analysis, hazard and operability study, failure mode and effects analysis);
- risk ranking;
- operability and maintainability reviews;
- code compliance reviews; and
- process safety management type reviews.

Chapter 6 identifies common water industry hazards and control measures that can be considered when conducting a hazard analysis as a part of the PtD methodology. Additionally, several important steps to take during a PtD analysis are as follows:

- Include HSE, O&M, and engineering staff to get a broad perspective on concerns.
- Review historical injury and incident data for common injury causes.
- Review government-mandated programs, such as permit-required confined-space entry or lockout/tagout, that require significant safety resources (staffing, special equipment and training, respirator and other PPE use) to meet the requirements.
- Look at existing facilities and processes and ask, "How could this facility or process be improved for safety?"
- Visit similar facilities at other water agencies and talk with them about their best safety management practices.

- If other agencies have already built or implemented a new technology, talk to them about what is working or not working for worker safety for that technology.
- Hazard identification should also be considered during the construction phase of a project.
- Consider new technologies such as BIM, IoT, AI and VR during the planning and design phase. Additional testing and studies are being implemented, benchmarking with utilities that have already implemented such technologies is encouraged.

Thinking through user and maintenance needs. Effective PtD must consider how workers will need to operate *and* maintain the equipment or process. Because the design engineer may not have operations experience, clear communication with O&M staff is necessary. O&M staff should clearly express their needs, including equipment access, space requirements to reach components for maintenance or dismantling, how equipment controls are to be calibrated, where workers will need to tie off, and other routine or common tasks. These needs concerning worker interaction with the process and equipment must be part of the data-gathering process early in the design phase because each interaction has potential worker safety consequences. The design engineer must ask questions that go beyond "What does the equipment or process need to do?" to determine "What does the worker need to do with, on, under, or on top of the equipment/process?"

Studies that support the prevention through design process. The design engineer can consider the consequences of potential hazards (affecting personnel, assets, the environment, and the Owners reputation) and issues affecting the continuity of operation and maintenance of the plant after construction and commissioning through the use of a detailed Hazard and Operability ("HAZOP") Study. The HAZOP Study should propose safeguards to prevent, control, or mitigate hazards and major operability problems. A HAZOP Study should be done for all plant processes, subsystems, utility systems (power, air, water, wastewater, etc.), pipelines, and associated equipment and control systems (including all operating modes and normal and abnormal operations such as commissioning, start-up, shutdown, cleaning, flushing, and emergency operation).

Similarly, the design engineer can identify, consider, and propose improvements to the design to address potential operational and maintenance issues (primarily usability, maintainability, operability, and health and safety) through the use of a detailed Access, Lifting, and Maintenance ("ALM") Study.

The HAZOP and ALM studies are based on the use of keywords using general arrangement drawings and piping and instrumentation diagrams to assess and identify any shortfalls or deficiencies in the design. Both studies should be completed near the start of the 60 percent detailed design.

Equipment specifications. The specifying and purchasing of equipment that meets safety and health goals is a PtD area with significant potential for success. Some examples include

- the Buy-Quiet approach adopted by the National Aeronautics and Space Administration (NASA) to specify noise emissions criteria along with other technical and performance criteria, to reduce equipment noise hazards (NASA 2014);
- specifying purchase of ergonomically adjustable chairs and workstations for control rooms with multiple users; and
- specifying standard fall-arrest and retrieval equipment for an organization so that equipment and controls are consistent between facilities.

Internal agency design standards. Developing internal agency design standards that both integrate lessons learned and provide a standard approach across the organization can simplify the process and yield consistent, successful results. These standards must meet at least minimum applicable regulatory standards, but then can include organization-specific criteria above and beyond the minimum requirements. Some examples include

- standard building roof designs (i.e., parapet, railing or tie-off anchor points);
- standard well or pump station designs (i.e., with no or non-permit confined spaces, low noise pumps, video feed, chemical secondary containment, chemical pump positioning to avoid splash, adequate ventilation, etc.);
- standard manhole designs incorporating both regulatory requirements and best practices (e.g., ladder restraint or confined-space entry enhancements);
- standard vault designs incorporating stairs, ventilation, and lighting to facilitate entry;
- standard tank roof designs for regular access (i.e., ladder/stair and railing) or emergency service access (tie-off anchor points); and
- standard chemical unloading pad and storage tank containment and detection systems.

Design process and proactive design safety review. Robust design review is another essential requirement for effective PtD methodologies. PtD considerations will begin with the alternatives analysis and conceptual engineering phases. At this early stage, only broad PtD assessments can be made, such as evaluations of different chemical process and siting options, or the identification of hazard categories that will need to be addressed further on in the design. In addition, any applicable, project-specific, internal design standards should be identified as part of the overall design criteria.

When the design is under way, project design review that includes safety concerns is critical at early, mid, and near-final design stages. The review should focus on

- whether or not the applicable safe design criteria have been incorporated in the design;
- resolving safe design issues that were identified as important but that could not be fully specified until more detailed design decisions were made; and
- identification of unanticipated concerns.

Engineering teams generally implement regular design review. For PtD review, the team must include individuals with health and safety expertise, as well as operations and maintenance expertise. Design review can be challenging if reviewers are not familiar with reading engineering drawings, if reviewers have trouble visualizing three-dimensional (3D) facilities from two-dimensional drawings, or if the design has many pages of drawings for a large, complex facility. Suggestions include having overview presentations from the design team, skills training on how to read engineering drawings, the use of design review checklists, and if feasible, the use of software programs that provide a 3D view of the facility. The final important step in safe design review is to have a good process to capture and respond to design review comments.

Constructability. The construction phase of a project may also provide opportunities for implementing PtD strategies that will keep construction laborers safe. Examples include the construction or assembly of modular roof or building units on the ground, then lifting the completed units into place, or the installation of fall-arrest system anchor points on structural elements before those elements are installed, so construction workers are able to use them during construction. PtD application during construction requires good knowledge of construction means and methods, which may necessitate input from construction management experts. Full implementation of PtD should include the construction phase.

Lessons learned. Every project is an opportunity to apply PtD concepts to eliminating, reducing, or controlling safety hazards. Every project is also an opportunity to identify additional safe measures and/or lessons learned, although some of those lessons may not be understood until workers have worked in the facility for a while. However, it is important for an organization to not only make necessary safety corrections, but to have a system in place to capture those lessons learned so that the information can be shared and considered during future project design.

REFERENCES AND BIBLIOGRAPHY

For more detailed information on the NIOSH PtD initiative begun in 2007, the initiatives of other professional societies, and regulations in some countries (e.g., United Kingdom), refer to the following.

American Institute of Chemical Engineers Center for Chemical Process Safety. 2008 (3rd ed.). *Guidelines for Hazard Evaluation Procedures*. New York: Wiley.

American Institute of Chemical Engineers. CCPS – Center for Chemical Process Safety | AIChE. www.aiche.org/ccps. Accessed June, 2022.

ANSI/ASSP Z590.3. 2021. Prevention Through Design Guidelines for Addressing Occupational Risks in Design and Redesign Processes. American Society of Safety Professionals.

General Services Administration. 2011. GIM Guide for Facility Management, Version 1.0 [Online.] Accessed December 16, 2021.

NASA Buy-Quiet Roadmap. A tool that supports the procurement of low-noise products. [Online.] Accessed December 16, 2021.

NIOSH (National Institute of Occupational Safety and Health). 2013. Prevention Through Design Initiative. Centers for Disease Control and Prevention. [Online.] Accessed December 15, 2021.

Water Research Foundation. 2010. *Water Utility Safety and Health: Review of Best Practices.* Report No. 3104. Denver, Colo.: Water Research Foundation.

Workplace Safety and Health Council. 2008. Guidelines on Design for Safety in Buildings and Structures. [Online.] Accessed December 16, 2021.

This page intentionally blank.

Chapter 5

Contractor Safety Management

Peter S. Puglionesi, David C. Baird, III

The water industry's construction and distribution system operations have the largest share of injuries of all water system operations (Figures 5-1, 5-2, and 5-3). Whether contractors are used only for major construction or for distribution system rehabilitation and repair, contractors have historically been responsible for, and the victims of, many major incidents. Incidents often include collapse of excavations, falls, being struck by vehicles and heavy equipment, being caught in rotating equipment, and release of hazardous energy.

Historically, many considered safety solely the contractor's responsibility, and some even mistakenly believed that liability increases when the owner improves contractor safety oversight. Beyond the human toll, water systems pay for contractor incidents in higher contractor overhead costs, distribution system and plant damages, service interruptions, and in many cases, damage to the organization's reputation. OSHA also holds owners accountable for lax oversight under its Multi-Employer Citation Policy (CPL 2-0.124 at: www.osha.gov/pls/oshaweb/owadisp.show_document?p_table=DIRECTIVES&p_id=2024). Robust contractor safety management programs have proven their value in reducing incidents, property damage, costs, and liability.

This chapter reviews established HSE practices that can be applied during contracting and construction to prevent injury to contractors, employees, and the public and prevent damage to property and operability. Some example forms have been provided in each section in this update to quickly allow organizations to implement the concepts outlined in this chapter.

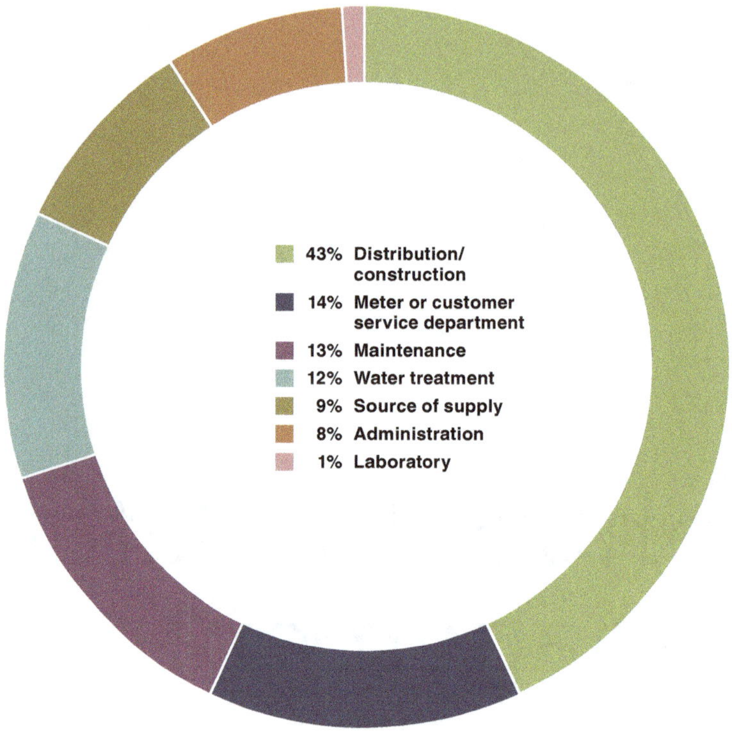

Source: Puglionesi 2003

Figure 5-1 Percent recordable cases with no lost workdays by type of operation (1999)

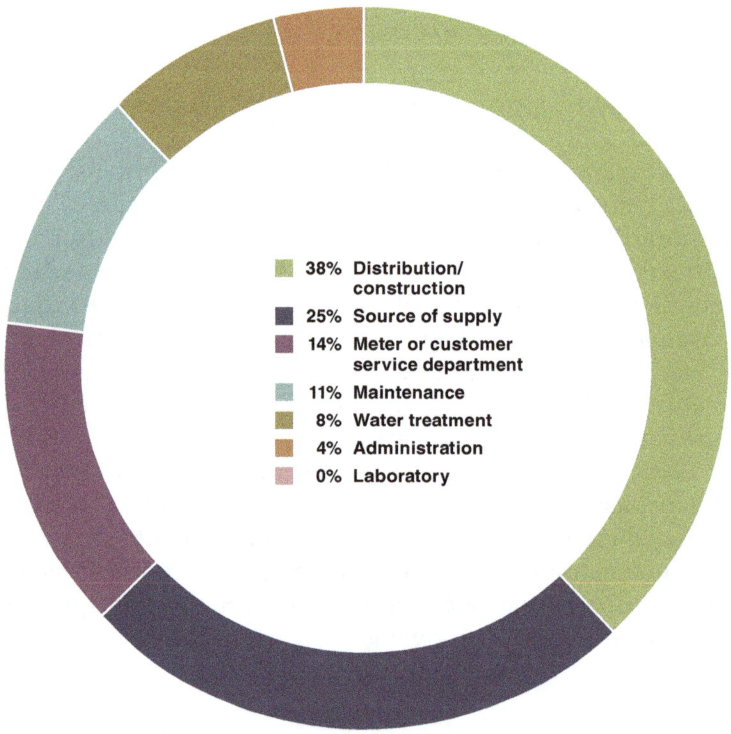

Source: Puglionesi 2003

Figure 5-2 Percent recordable cases with lost workdays by type of operation (1999)

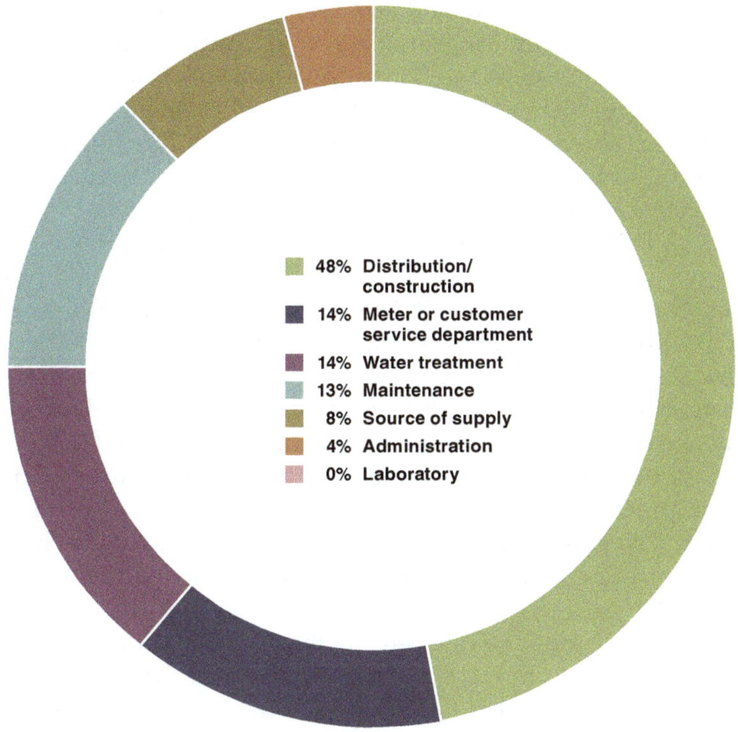

Source: Puglionesi 2003

Figure 5-3 Percent of lost workdays by type of operation (1999)

MAKING SAFETY PART OF THE CONTRACT

Organizations around the world have learned that they can improve contractor safety performance by comprehensively communicating and applying safety management requirements at all phases of the work, from before contractor selection through project completion.

Contractors will do what is necessary to win and hold a utility's business. In order to make safety performance and practices important to them, make it clear that it is important enough that contractors with poor safety records will be rejected, and insist on good contractor safety programs and practices that include identifying and correcting safety problems on an ongoing basis.

Successfully completing any project without an accident or safety violation depends upon both a comprehensive contractor safety management program and knowledgeable, engaged employees involved in contracts, purchasing, project management, and safety oversight.

Regulatory Requirements for Owners

While there are no generally applicable requirements for owners to have contractor safety management programs or for contractors to have comprehensive HSE programs, there are several relevant regulatory requirements and standards. Facilities that use chlorine, ammonia, or aqueous ammonia may be regulated under OSHA Process Safety Management (PSM) or US Environmental Protection Agency (USEPA) Risk Management Program (RMP) regulations (29 CFR 1910.119 and 40 CFR 68). These regulations were modeled after best practices developed for PSM in the petrochemical industry. Both regulations require contractor safety management. Both agencies can also enforce the "General Duty Clause" of the enabling legislation that requires employers to identify the hazards of any Extremely Hazardous Substance (under Emergency Planning and Community Right-to-Know Act regulations), design and maintain a safe facility to prevent releases, and minimize the consequences of releases which do occur.

OSHA PSM "applies to contractors performing maintenance or repair, turnaround, major renovation, or specialty work on or adjacent to a covered process. It does not apply to contractors providing incidental services which do not influence process safety, such as janitorial work, food and drink services, laundry, delivery or other supply services."

The standard requires the owner or operator, when selecting a contractor, to evaluate the contract owner or operator's safety performance and programs and periodically evaluate the contractor's performance during the course of the job.

In addition, the owner/operator of a facility is required to develop and implement safe work practices to control the entrance, presence, and exit of the contract owner or operator and contract employees in process areas and to notify the contractor of all known potential fire, explosion, or toxic release hazards related to the contractor's work and the process.

In turn, the contractor must ensure that each of his or her employees is trained in safe work practices for performing the job at hand and that the employees are instructed in the known potential fire, explosion, or toxic release hazards related to his/her job and the process, and the applicable provisions of the emergency action plan.

Documentation that each contract employee has received and understood the training is also required under the OSHA standard. The records must contain the identity of the contract employee, the date of training, and the means used to verify that the employee understood the training.

The standard also has notification requirements should any unique hazards occur because of the contract owner or operator's work, or of any hazards are found by the contract owner or operator's work.

Public Utility Commission (PUC) oversight is increasing in the areas of safety performance and safety management. For instance, in 2016, California Passed Senate Bill No. 62 to "establish within the commission the Office of the Safety Advocate to advocate for the continuous, cost-effective improvement of the safety management and safety performance of public utilities. The bill would require the office to promote public utility safety by: (1) advocating for effective public utility safety management and infrastructure improvements and for the transparency of safety information; (2) recommending improvements to the commission's safety management policy and procedures and its safety culture; and (3) informing the official record on safety-related risks in applicable commission proceedings and assisting the commission in its efforts to hold public utilities accountable for their safe operation." This Office of the Safety Advocate (OSA) specifically is focused on continuous and cost-effective improvement of the safety management and safety performance of public utilities which would include even contractor work. California is often a leader in this type of legislation, so it is possible that other states will begin to follow suit as public pressure from environmental and safety incidents continues to grow.

Even where there is no explicit contractor safety oversight provisions of state enabling legislation, PUC's have often exercised enforcement discretion over contractor safety oversight based on their core public safety and reliability requirements when serious contractor accidents occur in distribution system work.

Regulatory Requirements and Best Practices for Contractors

OSHA regulations do not require written comprehensive health and safety programs for the construction industry, although some specific regulations call for written programs or plans. For example, hazardous waste cleanup and emergency response operations regulations (29 CFR 1910.120) require the development of both a company program and a site-specific health and safety plan (HASP). The HASP addresses hazards that are specific to the conditions at the site and the work being done.

California construction safety regulations (Title 8 CCR 1509 and 3203) specifically require developing and implementing an effective construction Injury and Illness Prevention Program (IIPP). This mandatory approach is also taken in Washington, Hawaii, Australia, and much of the European Union. Regulations covering some industries or encouraging IIPP development exist in a total of 34 states. For example, California regulations require

- every employer to establish, implement and maintain an effective Injury and Illness Prevention Program;
- every employer to adopt a written Code of Safety Practices which relates to the employer's operations;
- the Code of Safe Practices to be posted at a conspicuous location at each job site office or provided to each supervisory employee who shall have it readily available;
- periodic meetings of supervisory employees under the direction of management for the discussion of safety problems and accidents that have occurred; and
- supervisory employees to conduct regular safety meetings with their crews at least every 10 working days.

The detailed IIPP regulations in Section 3203 are lengthy, but generally include

- identifying the persons with authority and responsibility for implementing the Program;
- training and instruction initially for changes in hazards, including supervisor training;
- a compliance assurance system, including employee recognition and disciplinary action;
- active two-way communication (meetings, posting, written communications, a means of anonymous employee hazard notification, and labor/management HSE committees);
- periodic inspection to identify/evaluate hazards, unsafe conditions, and work practices;
- procedures to investigate occupational injury or occupational illness; and
- procedures for timely corrective action.

There are two well-established industry consensus standards establishing best practices for IIPPs:

- ANSI/ASSP Z10 – 2012 Occupational Safety and Health Management Systems (American National Standards Institute/American Industrial Hygiene Association)
- ANSI/ASSP/ISO 45001-2018 Occupational Health and Safety Management Systems - Requirements with Guidance for Use (replaces OHSAS 18001 – 2007)

Contract Requirements

The legal foundation for enforcing an organization's program resides in the construction contract and construction specifications. As a program evolves, make sure that standard contract language supports the actions that may be necessary under a specific program. If not supported by the contract and specifications, contractors may request additional compensation for compliance and even legally challenge actions taken by the utility or its representatives. Depending upon the program, critical provisions of the contract and the specifications may include

- safety/environmental performance and/or program submittals required for either pre-qualification or selection;
- safety and environmental selection/exclusion criteria;
- safety program/project-specific plan submittals (e.g., health and safety plan, emergency/fire protection plan, job hazard analysis, hazardous material SDSs);
- pre-job orientation and planning meetings;
- site safety orientation for all contractor/subcontractor personnel;
- license/qualification/training requirements;
- contractor safety representative staffing/qualifications;
- contractor safety field inspection, self-assessment, and reporting;
- reporting/investigation of incidents and violations of laws or regulations;
- corrective action;
- contractor warning and suspension for failure to comply with safety requirements and grounds for termination; and
- contractor evaluation and its use in future contractor selection.

CONTRACTOR PREQUALIFICATION AND SELECTION

Selecting only contractors with an adequate safety record and safety program improves safety performance by

- eliminating the worst performing contractors, often those with whom staff would prefer not to work because of safety concerns; and
- encouraging all current and prospective contractors to improve safety performance; they know it matters to the utility and they may lose the business if they do not perform well.

For minor routine contracted services (e.g., inspection, calibration, testing, equipment rebuilding), this involves administering a periodic safety information request. For major or competitively bid projects, the safety information request becomes part of the bidding and selection process.

Organizations that are reluctant to prequalify contractors often believe that fair bidding laws restrict their ability to prequalify. This does not always prove to be the case, but the safety information request can be made after the presumed lowest qualified bid is identified, if necessary. The submittal is then evaluated against the contractor selection criteria to determine whether to prequalify or select the contractor.

Prequalification and selection can range from a simple review of performance data to a comprehensive review of both performance and safety programs. The simplest approach is to request the company's workers' compensation Experience Modification Rating (EMR) issued by the company's insurer for the last 3 to 5 years. An EMR of 1.0 is the average for the industry, 0.5 is half the industry average, and 2.0 is twice the industry average.

Table 5-1 Examples of prequalification and selection information to be collected from the contractor

Type of Information	Examples
Safety performance	• Experience Modification Rating (EMR) • Recordable injury/illness information (e.g., OSHA 300A summaries) • Number of hours worked; number of OSHA-recordable injuries/illnesses; number of fatalities; number of lost workday (LWD) cases • Recordable or LWD Incidence Rates (No. recordable or LWD cases × 200,000/employee hours worked) • List of OSHA notices of violation and fines • Safety/health-related judgments, claims, contract terminations, or pending/outstanding lawsuits
HSE programs	• Certification that contractor follows all applicable regulatory requirements • Description of the contractor's HSE programs • Copy of the HSE program • Description of the contractor's HSE training (e.g., course listing, frequency, syllabus)
Environmental performance	• List of environmental citations • List of reportable petroleum or chemical releases
Licenses/Certifications	• Copy of licenses, certifications, training, or experience documentation (e.g., crane, OSHA 10/30 training)
Experience/References	• Projects and references for similar scopes of work

More extensive safety information submittals may include any of the examples outlined in Table 5-1. (See also Figure 5-4, CS-01 – Contractor Prequalification Form.)

Some programs go beyond these practices to include specific safety practice questions, such as the following:

- Does the firm have a written safety and health policy or mission statement?
- Does the firm have a written safety and health training program?
- Does the company training program include the following?

Company HSE policies/programs	Lockout/tagout
New employee orientation training	Hot work
Emergency action plan	Electrical safety
Fire prevention plan	Bloodborne pathogens
Confined space entry	Scaffold
Excavation safety	Hazard communication
Fall protection	Hearing conservation
Personal protective equipment	Injury/illness recordkeeping
Respiratory protection	

- Is all HSE training documented? Is understanding/proficiency documented?
- What additional HSE training do supervisory personnel receive (foremen, crew supervisors, etc.)? What are the site safety representative's qualifications?
- Does the company have a new employee orientation and HSE training program?

- Are safety meetings regularly conducted?
  - How often?
  - Is there an established agenda?
  - Who ensures these meetings are conducted?
  - Are all employees required to attend?
- Does the company have a written disciplinary action policy on safety violations?
- To what extent does the firm use professional HSE staff?
- Are HSE inspections/audits conducted and documented?
- How are identified conditions brought to closure? Are corrected conditions documented?
- Is a Job Safety Analysis conducted? If so, when and for what activities?
- Is there a behavioral safety program? Behavior-based safety is a concept that accidents are a result of human error. Behavior-based programs address the cause of the errors (e.g., attitude, environment)
- Is there a "stop work" program for serious safety conditions? (Stop work programs require work to be stopped when a serious hazard is not addressed.)
- Is incident investigation conducted for all injuries? For near misses?
- Is safety a criteria in evaluating the performance of foreman, supervisors, management?

Contractor Acceptance Criteria

Some organizations are uncomfortable with disqualifying bidders based on their safety record, but this is now a common practice in industry and government. Selection criteria range from incidence limits to qualitative evaluations, to complex numerical scoring assigned to both performance and answers, to a series of questions. Some examples of simple contractor acceptance criteria are provided in Table 5-2.

Table 5-2 Examples of prequalification and selection criteria to be used by the owner/utility

	Acceptable	Corrective Action Required	Unacceptable
Experience Modification Rating (EMR)			
New	< 1.25	1.25 to 1.50	> 1.50
Improved	< 1.00	1.00 to 1.25	> 1.25
Mature	< 0.75	0.75 to 1.00	> 1.00
Advanced	< 0.50	0.50 to 1.00	> 1.00
Prequalification and Selection Criteria			
	No pattern of violations, fatalities, or serious injuries	Some history of violations, fatalities, or serious injuries	Pattern of serious violations, fatalities, or serious injuries
	Fully understands HSE program and training requirements (based on job and submitted information)	Some HSE program and training deficiencies or gaps (based on job and submitted information)	No or seriously deficient HSE program and training
	Has excellent qualifications and relevant experience	Has limited qualifications and relevant experience	No relevant qualifications and experience
	Has necessary licenses and certifications	Has some but not all necessary licenses and certifications	Does not have necessary licenses and certifications

Many organizations are initially worried that adopting definitive numerical criteria will exclude trusted contractors. A preliminary survey usually allays these fears because the vast majority of contractors have EMRs below 1.5. The EMR criteria can be used as a tool to gradually improve contractor safety performance, beginning at higher EMR thresholds, and gradually reducing them as the program matures and contractor performance improves. Some organizations have achieved a reduction in contractor EMRs to an average well below 0.5 over decades of applying a rigorous contractor safety management program.

There is no correct set of criteria for evaluating a contractor's safety record or program. Each organization should select the approach that best suits the organization and adjust the criteria with experience and to further encourage safety performance improvements among their contractors. Some organizations may want to avoid a hard exclusion criterion and always give its health and safety organization the ability to determine whether extenuating circumstances allow for acceptance under a corrective action program (e.g., a small company's rate may be greatly affected by such things as injuries due to a vehicle accident in transit to a work site).

Many organizations avoid an evaluation of the contractor's safety program because they believe that this is not their role, it is resource-intensive, and they might incur some liability for approving the contractor's program. To address this concern while still identifying weak contractor programs, water systems may decide to review contractor HASPs to identify deficiencies or gaps. Project engineering/management, purchasing and perhaps even the legal department should be involved in this decision making.

Requirements generally flow down to subcontractors for such things as insurance and compliance, but practices vary regarding flow down of safety performance selection criteria. An organization may decide to flow down all requirements to all subcontractors, apply them only to major subcontractors over a designated dollar threshold, identify a limited set of subcontractor criteria to all subcontractors (e.g., EMR) or apply none at all. Regardless, contractor and subcontractor safety requirements, along with any applicable flow-down clauses, should be explicitly written into the contract documents.

PRE-JOB CONTRACTOR ORIENTATION AND PLANNING

Prior to the commencement of work, an orientation meeting for the contractor's site manager/supervisor and safety representative should be conducted to ensure the contractor is fully prepared to comply with the safety requirements of the contract. The meeting will generally

- review the work, including key mobilization steps, heavy equipment, and projected effects on facility operations;
- review major facility petroleum/chemical hazards and the hazardous substances that will be brought on site to complete the work;
- confirm that the contractor understands its obligations to provide a site safety representative responsible for construction safety whenever work is conducted;
- share key facility emergency information, contractor work rules, and facility health and safety programs (e.g., lockout/tagout, hot work/safe work permits) that must be communicated to all contractor and subcontractor personnel before they begin work;
- confirm that the contractor understands its obligations to provide site safety orientation for all site workers (unless the facility conducts the orientation itself) and any other safety and skill training required for the safe completion of the work; and

- review key contract safety requirements (e.g., safety meetings, inspection, reporting, corrective action, discipline, and work suspension/termination provisions).

A contractor orientation checklist may be used to guide the pre-job orientation. A checklist also serves to document the transmittal of critical information and can include a contractor's acknowledgement that they understand the safety requirements (Figures A-2 and A-3.)

Some facilities prefer to provide a safety orientation to each and every contractor and subcontractor employee as a condition of gaining access to the facility to ensure that the orientation is rigorous and to ensure that there is documentation of the orientation and that each person understood the content. if this is made a contractor responsibility, the program should include a written acknowledgement by the contractor that they will provide appropriate site safety training to all contractor employees and subcontractor employees before working on the premises and transmit documentation that the orientations were conducted and understood (see Figure A-4).

CONTRACTOR WORK RULES AND GUIDANCE

The owner and contractors each have their own HSE programs. The owner should define when a contractor may operate solely under its own program and when it must also meet the owner's requirements. Generally, the owner's requirements are also applied when work is within the boundary limits of an existing facility or operation where the contractor may affect the safety of facility personnel and where an incident can damage property or the ability to operate. Good practice includes preparing and communicating the following to contractors:

- A brief list of contractor safety rules. This can also be used as handout for the contractor's employees and subcontractors.
- Key aspects of the facility emergency action plan (or other emergency plan), at minimum covering the facility's major hazards, emergency alarm system, evacuation routes, and assembly areas.
- The water system's own HSE programs, when the contractor is working on an existing operation where the owner's employees or facility could be affected by the work (e.g., lockout/tagout program, hot work/safe work permit,).

Large organizations sometimes develop procedures, guidelines, and/or model plans specifically for contractor use on their projects. (See Figure A-3, Form CS-03 – Facility Visitor and Contractor Orientation/EHS Requirements.)

Contractor Programs, Plans, and Training

As discussed previously, federal OSHA regulations do not require written comprehensive health and safety programs for the construction industry, although some specific regulations (e.g., PSM) call for written programs or plans and some states and countries require an IIPP.

While studies have shown that an IIPP requirement significantly reduces injury and illness rates (OSHA 2012), an organization may not feel it is appropriate to require all contractors to develop an IIPP in order to qualify to perform work. A safe construction environment can be maintained by some combination of company health and safety programs and the development and application of site-specific HASP requirements. Site-specific plans may include

- contractor's HSE policy and signed commitment;

- clearly assigned responsibility and accountability for site HSE;
- emergency or contingency plan, including contacts and evacuation procedures;
- fire prevention plan addressing hazardous material storage and use;
- qualifications and training of site safety representatives;
- description of major activities that affect hazards and safety programs which apply;
- licenses, certifications, and specialized training required based on the planned activities and equipment;
- communication of hazards and hazardous substances;
- periodic inspection to identify and evaluate hazards, unsafe conditions, and work practices;
- employee recognition and disciplinary action programs;
- immediate reporting of occupational injuries and illnesses;
- injury/illness/near-miss investigation procedure;
- procedures for timely corrective action;
- health and safety programs and/or JHA that address the anticipated activities and hazards and identify needed precautions and protective equipment;
- health and safety site orientation;
- training, tailgate meetings, safety meetings; and
- record keeping and submittals during the work.

If the above HASP approach is used, it should be flexibly administered to allow contractors with mature, effective health and safety programs to rely on and use their existing programs.

INJURY/ILLNESS/INCIDENT REPORTING AND INVESTIGATION

Many have learned in their own operations that immediate and accurate reporting of injuries and illnesses is essential for proper care, reporting, and case management and that the use of an occupational health provider (OHP) to rapidly assess and care for injuries is an effective tool for minimizing unnecessary hospital visits and vague or inaccurate diagnoses (see chapter 3). Prompt incident investigation and corrective action is essential to preventing reoccurrences. Late and inaccurate reporting of injuries and illnesses can be costly and complicated.

Many contractors, particularly small companies, are less experienced and less equipped to effectively manage injuries and are more likely to mismanage such cases. Because the utility staff could potentially be involved in cases of serious injury and since the cost is ultimately built into the contractor's overhead, it is essential that contractors are required to promptly report and investigate injuries and illnesses. At a minimum, contractors should immediately report incidents and provide a copy of the investigation report. Some owners also participate in the investigation or conduct their own independent review. Organizations with aggressive contractor safety programs have even required immediate consultation with their own OHP.

Owners at OSHA PSM-regulated facilities are required to maintain a log of contractor injuries and illnesses. This can be a log that they maintain or a copy of the contractors' logs. Beyond the requirements, consider copying the contractors' performance data logs to measure the effectiveness of the contractor safety management program.

CONTRACTOR EVALUATION

Evaluating contractor safety and health performance provides important feedback to the contractor and helps determine their consideration for future work. Monitoring and evaluating performance can be done in a variety of ways. For small projects, informal site supervisor and operator monitoring may be the most efficient and effective choice to assure safety compliance. For larger projects, approaches vary from integrating this responsibility with the construction/project manager's other duties, to using health and safety staff, to using outside resources (consultants or insurance company representatives).

An evaluation form can facilitate consistent review practices and documentation. If the organization decides to perform active, frequent auditing of safety performance during the project, a detailed checklist of common construction safety requirements/issues is helpful. This checklist is usually different from an overall safety performance evaluation, which is used to support future contractor selection decisions. Detailed auditing may identify hazard and noncompliance issues and documentation should include corrective actions taken by the contractor. (See Figure A-5, CS-05 – Contractor Safety Evaluation.)

For OSHA PSM facilities, formal evaluation of the contractor's performance in meeting their obligations at least once during the work is required. This can be performed by the person who is assigned to manage the contractor or to provide safety oversight. The scope and frequency of the evaluation is discretionary.

In any addition to safety and performance evaluations conducted during a project, a thorough post-work evaluation with metrics for contractor requalification should be considered. In addition to evaluation of their safety performance, regulatory compliance and lagging indicators (e.g., injury and incident rates) on the project, the post-work evaluation could consider additional factors, such as competence of employees; quality of work; communication/coordination with host employer; and promptness in meeting deadlines. Contractors that struggle to find quality staff or consistently miss deadlines may be tempted to compromise safety to increase productivity or overcome poor work quality issues. Having a post-work evaluation can help identify the very best overall contractors.

Special Hazards of Construction

Health and safety issues and hazards applicable to construction are similar to those encountered used in water system operation, but certain hazards and issues have greater significance and require greater emphasis. These include

- cranes, hoists and rigging—dropping, toppling, collapse;
- excavation—underground lines, adjacent heavy equipment, danger of collapse;
- confined space—physical, chemical, oxygen, hot work;
- fall protection—falls from leading-edge work, inadequately guarded edges, worker interaction with lifts;
- hot work—fire, inadvertent work on active or inadequately cleared equipment;
- lockout/tagout—inadequate lockout/individual protection, inadvertent work on active or inadequately cleared equipment;
- contractor work in or around high hazard chemicals or processes;
- traffic—struck by vehicle; inadequate warning and transition zones;
- security—vulnerabilities due to inadequate identification requirements and access control at construction entrances; and
- emergencies—inadequate personnel accounting and emergency access.

Pre-startup Safety Review

Chapter 4, Prevention Through Design, addressed techniques for improving safety both during construction and during the operating life of the facility. Another key element of safety management is reviewing the completed facility during the commissioning phase to ensure that the new or modified facility is ready to safely start up and operate. Pre-startup safety review generally includes

- inspection for hazards and compliance with standards, often using a checklist;
- verification that construction and equipment is in accordance with design specifications;
- verification that adequate safety, operating, maintenance, and emergency procedures are in place;
- verification that prior design/construction review recommendations have been resolved or implemented;
- completion of operator training on new or improved equipment, processes, and materials; and
- establishment of maintenance schedules and replacement part requirements.

Fostering Partnerships

Partnerships can be an effective means of encouraging participation and success in improving safety performance. Partnerships can be sought with frequently used contractors, with regional contractor associations, and with a regional industry consortium. The regional labor/industry consortium approach has been successfully used to promote training and basic understanding of minimum safety requirements and practices. The Delaware Valley Petroleum Refiners Association partnership with regional nonprofit and union organizations provides process safety management curriculum for employees working in refineries and serves as a good model for other industries. The program is an eight-hour training session followed by a written test (IUOE 2008).

THIRD-PARTY CONTRACTOR SAFETY MANAGEMENT

Many organizations use some form of third-party contractor safety management support. This ranges from tasking a third party with selected tasks such as obtaining insurance certificates to spot safety inspection to full-time presence at major project sites to employing a third party for comprehensive contractor safety program management. Independent construction oversight can be included as a contractor obligation, integrated into a construction management contract, or contracted directly by the utility.

The National Safety Council (NSC) published a study in 2017 titled "Making the Case for Contractor Management: Examining the Safety Benefits of Third-Party Management." This report summarized the research on the effectiveness of using third-party contractor management potentially resulting in better safety performance for host employer contractors. One result of this study was that by industry, contractors in the third-party data set did have better lagging indicators in average TRR (Total Recordable Rate), DART (Days Away. Restricted, Transferred Rate), and LWR (Lost Work Rate) than industry as a whole and better than the cross-industry averages including construction. While this was a specific third party provider data subset, it was independently analyzed by NSC and it does provide a data point that demonstrates the value of a formal process of prequalification and ongoing evaluation. This performance could no doubt be replicated by a thorough in-house contractor safety management system since the comparison is to the balance of the utility industry, including those without a strong contractor safety management system and culture. For organizations that lack sufficient staff knowledge, a comprehensive contractor management system and a strong contractor safety oversight culture—third party management may be the fastest route to contractor safety improvement. This section does not promote or endorse the use of contractor safety programs. See Appendix A for examples of forms that can be used as resources.

REFERENCES AND BIBLIOGRAPHY

California Code of Regulations. Title 8, Section 1509. Injury and Illness Prevention Program. Subchapter 4. Construction Safety Orders Article 3. General. https://www.dir.ca.gov/title8/1509.html. Accessed December 16, 2021.

California Code of Regulations. Title 8, Section 3203. Injury and Illness Prevention Program Subchapter 7. General Industry Safety Orders. Group 1. General Physical Conditions and Structures Orders. https://www.dir.ca.gov/title8/3203.html. Accessed December 16, 2021.

OSHA Injury and Illness Prevention Programs White Paper. 2012. www.osha.gov/dsg/topics/safetyhealth/OSHAwhite-paper-january2012sm.pdf. Accessed January 8, 2014.

Occupational Safety and Health Administration. 2004. *Code of Federal Regulations*. Title 29, Subpart H, Hazardous Materials, §1910.119, Process safety management of highly hazardous chemicals.

Occupational Safety and Health Administration. 2004. *Code of Federal Regulations*. Title 29, Subpart H, Hazardous Materials, §1910.120, Hazardous waste operations and emergency response.

Puglionesi, P. 2003. "Identifying and Prioritizing Emerging Safety Issues in the Water Industry," AWWA Research Foundation Project 2684. Denver, Colo.: Water Research Foundation.

SB62 | California 2015-2016 | Public Utilities Commission: Office of the Safety Advocate. | TrackBill.

USEPA. 1994. *Code of Federal Regulations*. Title 40, Protection of Environment Chemical Accident Prevention Provisions, § 68.

Table A-1 Example contractor safety program process forms

Facility	1. Initiates process by sending Contractor Safety Packet with the following forms to contractor contact: • CS-01 – Contractor Prequalification Form • CS-02 – Contractor Orientation Checklist • CS-03 – Contractor Site Safety Information • CS-04 – Contractor Employee Training Record • CS-05 – Contractor Safety Evaluation
Contractor	2. Submit completed *CS-01 – Contractor Prequalification Form* and Supporting Documents to Facility
Facility	3. Review Contractor Prequalification Submittal and Approve/Deny
Facility / Contractor	4. If approved, communicate facility hazards and safety requirements (*CS-03 – Contractor Site Safety Information*) and document this communication on form *CS-02-Contractor Orientation Checklist*
Facility	5. Provide copy of completed *CS-02 – Contractor Orientation Checklist* to Contractor.
Contractor	6. Train contractor employees and subcontractor employees on information provided in Contractors Safety Packet. Document employee training on form *CS-04 – Contractor Employee Training Record*
Facility	7. Confirm all contractor employees provide a completed *CS-04 – Contractor Employee Training Record* before starting work.
Contractor	8. Perform work in accordance with project specifications, facility requirements and applicable federal, state, and local codes and regulations. Report Injuries, Illnesses, Incidents and Near Misses to the *Facility Representative*.
Facility	9. Periodically Evaluate Contractors Performance and document on *CS-05 – Contractor Safety Evaluation*.

GENERAL INFORMATION		
Company Name/Address:	Phone: Fax:	Email (person completing form):
Safety Contact:	Title:	Phone:
Years in business under the present firm name:		
Insurance Meeting Site Requirements: ❑ Yes ❑ No (attach explanation)		❑ Attach Certificate
Location and Description of Work for this Facility (or attach work scope summary):		
Major Jobs Completed in Last 3 Years:		
Customer/Location	Type of Work	Contact for Reference/Phone

SAFETY PERFORMANCE INFORMATION

OSHA citations or safety/health related judgments, claims, contract terminations or pending/outstanding lawsuits (including by your or customer employees) against your Company in the last 5 years? ❑ Yes ❑ No
If Yes, attach OSHA citations and explanation of any other items.

Provide injury/illness data for the past 5 years:	Year:	Year:	Year:	Year:	Year:
Workers Comp Experience Modification Rate (EMR)* State: __________ Anniversary Date: ________					
Employee hours worked (excluding subcontractors)					
Total # of OSHA Recordable Injuries and/or Illnesses					
# of Injury or Illness Related Fatalities					
# of Lost Workday (LWD) Cases (injuries/illnesses with days away from work, or days of restricted work)					
Lost Work Day Incidence Rate (# LWD Casesx200000/Employee Hours Worked)					

SAFETY PROGRAM INFORMATION		
Do you have written Safety and Health Programs? ❑ Yes ❑ No ❑ Attach list of programs (e.g., Lockout/Tagout, Hazcom, PPE, Confined Space Entry, etc.) or Table of Contents.		
Do you have a training program to support the above Programs? ❑ Yes ❑ No ❑ Attach list of training provided to employees who will work at the facility.		
Do you hold safety meetings? ❑ Yes ❑ No	Frequency of safety meetings:	Are safety meetings documented? ❑ Yes ❑ No
Do you have experience working on or around process systems that contain chlorine, chlorine dioxide solution, ammonia or any other material listed in 29 CFR §1910.119, Appendix A and 40 CFR § 68.130? ❑ Yes ❑ No		

Certification: I certify that the Health and Safety and Training Programs for our company comply with OSHA and other applicable regulatory requirements and that the information presented on this form and attachments is accurate and complete. Our company is aware and understands its obligations under the OSHA and other applicable standards and will abide by applicable facility Contractor Safety, Health & Environmental procedures, as appropriate. Our company will advise the facility prior to beginning work, of any hazardous substances it or any of its sub-contractors will bring on premises and/or any safety issues that could be created by its work. Such information, including MSDS for hazardous substances, will be communicated in writing to the Facility Representative. Likewise, if the Contractor's work will pose any safety issues for the Facility's employees, the issues will be discussed with the Facility Representative before commencing that portion of the work.

Signature: ____________________________ Title: ___________________________

Print Name: ____________________________ Phone #: _________________ Date: ____________

Facility Determination: ❑ Met requirements ❑ Disqualified ❑ Accepted Corrective Action Plan

Name: ______________________ Signature: ________________________ Date: ____________

* Attach insurer correspondence with latest year EMR data.

Source: Applied EHS Management, Inc.

Figure A-1 CS-01 – Contractor pre-qualification information

CS-02 - Contractor Safety Orientation Checklist

(Orientation for Contractor/Subcontractor Project Managers and Site Representatives at Start of Work)

Contractor Company Name: ______________________ *Date: ________

* Review this information with this Contractor within 1 year from this date (or sooner at the discretion of Project Manager and/or Facility Representative)

Facility: ______________________

The Project Manager or designee will review and initial each of the items below that apply to the specific Contract work before start of any contract work. The Contractor Representative and Project Manager shall sign this Checklist indicating that they have reviewed their responsibilities for these initialed items as they are outlined in the Contractor Safety Procedure.

Initial

_______ Provided and reviewed a copy of the Contractor Safety Procedure and attached Contractor H&S Requirements Summary

_______ Personal Protection Equipment (reviewed site requirements and Contractor specified PPE)

_______ Provided and reviewed the following FACILITY safety programs applicable to the work (check all that apply and add all other applicable programs provided) and discussed coordination between Contractor Programs and Facility Programs. Discussed Contractor's responsibility to ensure that all contractor employees are trained in the necessary safe work practices.:

- ____ Lockout/Tagout Program — Other (list): ____________
- ____ Confined Space Program — ____________
- ____ Hot-Work Program — ____________
- ____ Excavation Safety — ____________

_______ Reviewed the following high hazard activities applicable to the work and key OSHA and FACILITY requirements (check all that apply and add all other relevant safe work practices):

- ____ Ladders/Scaffolds/Man Lifts — Other (list): ____________
- ____ Fire Safety — ____________
- ____ Electrical Safety — ____________
- ____ Respiratory Protection — ____________

_______ Reviewed access control, including daily facility sign-in/sign-out and not entering restricted high-hazard process areas (e.g., hazardous chemical storage) unless authorized.

_______ Reviewed location of restricted high-hazard chemical process areas and other hazardous areas.

_______ Facility Emergency Evacuation Procedures/Plans/Alarms/Routes (provide evacuation map)

_______ Injury and Illness Reporting

_______ Chemicals/Hazard Communication (list of project chemicals, quantities, and SDSs provided)

_______ Waste/Chemicals Handling Procedures (for oil/fuel/chem. to be used & wastes generated)

_______ Other: Please Describe ______________________

CONTRACTOR CERTIFICATION: I fully understand the above information and certify that:

- I will notify the ***Facility Representative*** as soon as I become aware of any unique hazards associated with our work, and will explaining those hazards and needed controls before introduction of the hazard.
- I understand my responsibility to immediately inform the ***Facility Representative*** of all chemical or petroleum releases, injuries or illnesses or any incident and near miss resulting from work performed at the facility.
- I understand my obligation as a Contractor to ensure that only my employees and my Subcontractor's employees who have successfully completed training on the above information, who are trained in the safe execution of their work, and who hold current, valid licenses and certifications as required, will work at the facility and that they abide by applicable FACILITY policies and procedures.

______________________ Contractor Representative/Date

______________________ Facility Representative/Date

Source: Applied EHS Management, Inc.

Figure A-2 CS-02 – Contractor safety orientation checklist

Contractors: All your employees and Subcontractors that will work at the facility must receive this Orientation before they work. Either request a Facility orientation session employees or provide documentation to the Facility Representative that you trained them on this information.

CONTRACTOR ORIENTATION

BASIC SITE INFORMATION:

- Contractors will not commence work until a contractor orientation has been conducted for the Contractor by **Facility** personnel. (See Contractor Orientation Checklist)
- Cameras are not allowed without authorization.
- Alcoholic beverages and controlled substances are not permitted on the property. Smoking is not permitted on the property.
- All Personnel are expected to conduct themselves in a manner to promote a workplace that is free from all forms of discrimination and harassment of any kind.

RESTRICTED ITEMS:

- Do not bring the following items on the property: Firearms, Ammunition, Weapons, Cameras (without authorization), pets, liquor or other controlled substances)
- Do not bring the following items into the plant: Food and Drink, jewelry or watches.

You SHALL NOT introduce any hazardous material to treatment equipment in contact with treated water without approved plans to remove residues to levels that will not affect water quality. Tampering is a violation of federal law.

RESTRICTED AREAS:

- Chlorine Storage Room

ACCESS CONTROL:

- Contractors are restricted from entering any production area unless access is required to perform their work and they are authorized for entry.
- Contractors must sign in and out daily on the entry log at front desk
- The Administrative Coordinator will confirm that contract employees coming on site are listed on the approved contractor list or provide a valid "Contractor Training Certification"

INJURIES/INCIDENTS

- Report all injuries/illnesses requiring First Aid, Medical Treatment, or Transport to a hospital and any Incident resulting in damage to **Facility** equipment or facilities to the ***Facility Representative*** as soon as possible.
- All significant near misses are required to be reported to the ***Facility Representative***.

CONTRACTOR EMERGENCY PROCEDURES:

In the event of an emergency evacuation, you will be notified by audible alarms, PA system announcement or by word of mouth. Upon hearing an alarm/notification to evacuate:

1. Turn off any equipment you are operating,
2. Evacuate the building through the nearest and safest emergency exit, and
3. Move to the facility assembly areas for a head count **(maps posted throughout facility):**

Contact the ***Facility Representative*** prior to calling 911, **unless there is a life-threatening injury.**

All Contractor/Sub employees must be accounted for in an emergency. Do not leave the assembly area until instructed to do so.

For chemical releases, observe the windsocks and move to the nearest cross-wind or upwind location. Do not travel through a release plume. Remember – going downwind will prolong your exposure.

CONTRACTOR SAFETY REQUIREMENTS

REGULATIONS & PROGRAMS:

- Contractors and their employees and subcontractors are required to follow all Federal, State and Local Safety, Health, and Environmental regulations. (e.g. OSHA, state agencies, USEPA).
- Contractors shall coordinate with the ***Facility Representative*** on all activities subject to **Facility** safe work practices to determine whether to follow **Facility** programs or their own OSHA compliant safety programs. These include, but are not limited to: Confined Space Entry; Hot Work; Lock Out/Tag Out (LO/TO); Waste Handling and Disposal

HAZARDOUS CHEMICALS (SDSs):

- This **Facility** uses certain hazardous chemicals in the plant. Contractors or Subcontractors will be notified of any hazardous chemicals in areas which you may be working. Safety Data Sheets (SDS) can be obtained from the SDS Binders located in the **safety office and control room**.
- Contractors or Subcontractors shall not bring any hazardous chemicals onto the property without prior approval of ***Facility Safety***. A current SDS must be furnished to **Facility Representative**.

Source: Applied EHS Management, Inc.

Figure A-3 CS-03 – Facility visitor and contractor orientation/EHS requirements

CONTRACTOR SAFETY REQUIREMENTS

HOUSEKEEPING

- Contractors shall maintain a safe and clean working environment (Practice Good House Keeping).
- Tools, equipment, supplies (chemical or otherwise) and wastes shall be removed from the work area when Contract Employees finish work for the day or (for longer duration work) ensure they are neatly staged and do not encumber **Facility** operations, create trip hazards, or block aisles/electrical panels/fire extinguishers/exit doors.

CONFINED SPACE ENTRY (CSE):

- Contractors shall notify the ***Facility Representative*** prior to any confined space entry.
- Contractors may request entry under the **Facility** program (after receiving training) or provide evidence of a compliant program and request authorization to enter under Contractor's own program.
- Contractors entering under their own CSE program shall be responsible for all monitoring, retrieval setup and coordination of rescue before entry.
- Contractors shall abide by **Facility** or their own OSHA compliant CSE Program and ensure that their employees/subs engaged in confined space entry are fully trained in the requirements of that program. **Facility** program and confined space classifications can be made available upon request.

HOT WORK:

- Contractors shall obtain a **Facility** Hot Work Permit for all welding/cutting/other hot work in active operating areas. Where the ***Maintenance Supervisor*** completely turns an area over to the Contractor, they shall abide by their own OSHA compliant Hot Work Permit program.
- Contractors shall ensure their employees doing Hot Work are trained in equipment use/Hot Work Permit program requirements.
- Permits shall remain at the worksite until the work is complete including clearance by the Fire Watch.
- Any welding and/or cutting operations require an additional person for fire watch that is equipped and trained for fire extinguisher use.

LOCKOUT / TAGOUT (LOTO):

- NEVER remove LOTO locks/tags/other hardware applied by others to prevent activating or opening equipment that can injure or kill by releasing hazardous energy (electricity, chemicals, pressure, spring pressure, weight drop, etc.).
- Contractors shall not perform work on equipment with hazardous energy unless authorized by **Facility Representative** to do so under the Lockout/Tagout (LOTO) program. The ***Facility Representative*** will provide contractors who are authorized to perform Lockout/Tagout (LOTO) under **Facility** LOTO Program with further orientation material.
- LOTO Work Affecting **Facility** Employees/Operations - Generally for work which affects/interferes with **Facility** employees/operations, **Facility** personnel must perform all such equipment shutdowns and lockouts prior to Contractor maintenance activities. Then each contractor employee trained/authorized to work on the locked-out equipment must apply their own locks to each energy source locked out to ensure that the hazardous energy remains controlled and the equipment cannot be operated until their locking device is removed. A copy of the **Facility** Lockout/Tagout program will be provided with detailed requirements.
- LOTO work Not Affecting **Facility** Employees/Operations - ***Facility Representative*** may authorize Contractor to work under their own LOTO Program when work does not affect or interfere with **Facility** employees/operations. Contractor shall notify **Facility Representative** of every LOTO/ and give the **Facility** the option of also applying their own shop lock(s) to the equipment to prevent re-start until the **Facility** lock(s) is removed. Contractor shall notify when returned to service.

Source: Applied EHS Management, Inc.

Figure A-3 CS-03 – Facility visitor and contractor orientation/EHS requirements (*continued*)

CONTRACTOR ENVIRONMENTAL REQUIREMENTS

SPILL PREVENTION AND CONTROL

- To minimize the risk of spills or releases to the environment, contractors must employ appropriate protective procedures such as secondary containment, employee training, overflow protection, and other measures as part of activities involving the use, storage, or handling of petroleum products or hazardous materials on our property.
- Containers of hazardous materials and petroleum products should be stored in order to prevent releases to the environment. This requires selecting locations and methods to minimize exposure to rainfall, surface water, and the ground.
- Enclosures, shelters, and secondary containment should be used where appropriate. Containment pans should be placed under equipment where there is the potential for a leak or discharge.
- Prior to discharge of a secondary containment to the storm water system an area that is exposed to rainfall:
  a. inspect the primary container for signs of leakage and
  b. inspect the containment system by visual observation for color, foam, outfall staining, visible sheens, and dry weather flow.
- The discharge of a containment system that has evidence of contamination is prohibited. The responsible contractor must maintain a log indicating the individual making the observations, description of accumulated storm water, and the date and time of release. Submit a copy of the log to the ***Facility Representative***.

PERSONAL PROTECTIVE EQUIPMENT (PPE)

PERSONAL PROTECTIVE EQUIPMENT (PPE):

Construction and Demolition Areas

- Hard Hat
- Safety Glasses
- Safety footwear meeting the ANSI Standard Z41-1999 requirements for toe-cap protection
- Gloves appropriate for the task when handling objects or substances that could cut, burn, injure the hand, or be absorbed into the skin, and when exposed to harmful temperature extremes.

Unless otherwise noted, contractors are to provide the required and needed personal protective equipment, medical clearance, and necessary equipment specific training. Employees must be trained on the use, inspection, care, and storage of personal protective equipment. An employee who refuses to use the prescribed personal protective equipment or willfully damages this equipment may be immediately removed from our premises. The contractor must furnish any equipment required by unusual circumstances (such as high temperature work or handling corrosive liquids) and not specifically covered in this section. Use of such must be reviewed with the ***Facility Representative.***

OSHA PROCESS SAFETY MANAGEMENT (PSM) & EPA RISK MANAGEMENT PROGRAM (RMP)

This **Facility** stores and uses chlorine in excess of the regulatory thresholds and is therefore required to have a Process Safety Management (PSM) Program under 29 CFR 1910.119 and a Risk Management Plan (RMP) under 40 CFR 68. These programs are available for review by contacting our ***Facility Representative***. **Chlorine, at our facility is used for disinfection.**

We want to ensure that you are aware of any potential danger that could occur while you are in the facility and how to respond in the event of a chlorine emergency.

Source: Applied EHS Management, Inc.

Figure A-3 CS-03 – Facility visitor and contractor orientation/EHS requirements (***continued***)

Chlorine Hazards	UN / NA 1017. Chlorine is a toxic, corrosive gas that attacks the mucous membrane of the eyes, nose, throat, and lungs. Liquefied chlorine vaporizes if released and will form a dense gas that can accumulate in low areas. Liquefied chlorine can cause frostbite and tissue damage; however, the corrosive properties of the liquid and chlorine gas is significantly more dangerous. Chlorine gas has a characteristically pungent and irritating odor. This sharp odor serves as a warning signal with even low concentrations of chlorine in air readily detectable below the safe exposure limit. While chlorine does not burn, it is an oxidizer. Containers should be separated from flammables/combustibles and protected from heat. Extreme caution must be used when responding to releases. Significant chlorine vapor thresholds (ppm by volume) are summarized below: • 0.3 – 0.5 ppm (0.9 – 1.5 mg/m3) - Odor Threshold • 0.5 ppm (1.5 mg/m3) - 8-Hour ACGIH TWA, no effect: • 1.0 ppm (2.9 mg/m3) - Don Respirator (if qualified) - 15-Minute OSHA CEL/ACGIH STEL: • <3 ppm (8.1 mg/m3) - 60 min. ERPG-2, No permanent injury • 10 ppm (29 mg/m3) - 30 min. IDLH, able to escape: • 15 - 30 ppm (44 - 87 mg/m3) - Pulmonary edema (fluid in lungs), respiratory arrest, and potential permanent injury • 100 - 1000 ppm (290 - 2900 mg/m3) - Potentially lethal, even for short exposure Contact the **Facility Representative** for a chlorine SDS
Chlorine First Aid	If exposure does occur, the victim should be removed to fresh air and medical attention should be obtained at once. More complete information on health hazards and first aid for chlorine exposures can be found in safety data sheets (SDSs). The following summarizes key first aid information found on the SDS: **Eye: Irrigate immediately**....If this chemical contacts the eyes, immediately wash (irrigate) the eyes with large amounts of water, occasionally lifting the lower and upper lids. Get medical attention immediately. **Skin: Water flush immediately**....If this chemical contacts the skin, immediately flush the contaminated skin with water. If this chemical penetrates the clothing, immediately remove the clothing and flush the skin with water. Use caution when removing clothing as it may be frozen to the skin. Get medical attention promptly. **Breathing: Respiratory support**....If a person breathes in this chemical, move the exposed person to fresh air at once. If breathing has stopped, perform artificial respiration. Administer oxygen if breathing is labored. Keep the affected person warm and at rest. Get medical attention immediately for any inhalation – health effects can be delayed. **Swallow: Medical attention immediately**....If this chemical has been swallowed, get medical attention immediately. Do not induce vomiting.

Source: Applied EHS Management, Inc.

Figure A-3 CS-03 – Facility visitor and contractor orientation/EHS requirements (***continued***)

This form is filled out by the **contractor employee** and **contractor supervisor/trainer** for every contract employee before work can begin. A copy of our Safety Policies and Site Safety Information are available to the contract employer.

Contractor Company Name: ______________________________

Contractor Employee Name: ______________________________

Training

Initial each applicable box when instruction is completed and understood:

1. Contractor Company Safety Policy and Program ☐
2. Facility Site Safety Information for Contractors ☐
3. Facility Safe Work Practices (LOTO, Hot Work, Electrical Safety) ☐
4. Personal Protective Equipment (PPE) Required ☐
5. EAP - Fire Prevention, Alarm awareness, Location of Exits & Assembly ☐
6. Known Potential and Special Hazards of the Job and Process ☐
7. Manufacturer's procedures & skills training applicable to the work ☐

What means were used to verify this training was understood?
(e.g., written quiz, verbal Q&A, visual observation of task performance and proficiency, etc.)

Contract Employee Certification

I reviewed the above checked items relating to the safety rules and safe work procedures for work

at: ______________________ (*Facility*) on ______________________ (*Training Date*).

By my signature, I certify that the information provided by the facility has been thoroughly read, understood, and will be adhered to.

______________________ *Contract Employee Signature* ______________________ *Date & Time*

Contract Supervisor / Trainer Certification

I certify that I have trained the employee listed above in the information provided by the facility. The employees listed above have also been trained in the procedures required to safely perform their work (including any skills training and manufacturer's procedures applicable to the work)

______________________ *Contract Employer/Trainer Signature* ______________________ *Date & Time*

Source: Applied EHS Management, Inc.

Figure A-4 CS-04 – Contractor employee training certificate

This form is to be completed by the Facility Project Manager or Safety Representative during Contractor work to Evaluate Contractor Performance – Minimum 1 Time/Year per Contractor

Contractor Company Name: ______________________________

Facility: ______________________________

Evaluated By/Date: ____________________ / ____________

1) *Unacceptable* (2) *Needs Improvement* (3) *Average* (4) *Good* (5) *Excellent*

1. Follows Facility Contractor Safety Program Requirements	1	2	3	4	5
2. All Contractor/Subcontractor Employee Trained (certificates on file)	1	2	3	4	5
3. Competence of Employees	1	2	3	4	5
4. Safety in Work / Follows Facility Safety Rules	1	2	3	4	5
5. Positive Attitude in Planning/Communicating/Completing Work	1	2	3	4	5
6. Quality of Work	1	2	3	4	5
7. Housekeeping of Contractor Work Areas	1	2	3	4	5
8. Communication/Coordination with Facility*	1	2	3	4	5
9. Ability to Follow Directions	1	2	3	4	5
10. Promptness in Meeting Deadlines	1	2	3	4	5
Add Up the Rating Numbers->					
Multiply by 2 to get Rating Score					
If any area is rated "1" or "2" then corrective action is required regardless of overall score.					

Rating Scores

Score	**Category**	**A contractor with this score exhibits:**
1-20	Unacceptable	Fails to meet minimum requirements, Serious deficiencies/hazards; or slow/no corrective action
21-40	Needs Improvement	Many minor deficiencies, a serious hazard observed; or slow corrective action
41-60	Average	Generally adhered to regulatory and facility requirements & promptly corrects minor deficiencies, improvement possible
61-80	Good	Consistently complies with regulatory and facility requirements.
81-100	Excellent	Always adhered to regulatory and facility requirements; self-policing, pro-active program

Comments/Corrective Actions Taken

*Includes communication of hazards, injuries, illnesses, incidents, near misses

Source: Applied EHS Management, Inc.

Figure A-5 CS-05 – Contractor safety evaluation

This page intentionally blank.

Chapter 6

Common Hazards and Safety Programs

Dr. Nicholas Ashbolt

This chapter provides a brief introduction to hazards that are commonly faced in the water industry. While not all-inclusive, this chapter does cover many of the hazards that managers, engineers, HSE staff, and operators will face and, they may want to consider when planning through design (PtD), developing Job Safety Analyses (JSA), or selecting contractors. This information was assembled by experienced water industry safety professionals based on their knowledge of each hazard, and it serves as a great tool to evaluate whether or not a particular hazard applies to a particular utility. Controlling these hazards may require further investigation of the resources provided and, in some cases, formal programs and training.

The hazards addressed in this chapter are organized categorically as follows:

Physical Hazards

Arc flash

Permit-required confined space entry

Excavation and Trenches

Hazardous energy control (Electrical Safety)

Hot work

Materials handling

Noise exposure

Chemical Hazards

Chemical exposure and highly hazardous chemicals

Compressed gases

Asbestos, lead, and mercury

Silica

Hazardous waste

Pesticides

Radionuclides

Pressure vessels

Tractor operation

Traffic and the work zone

Driving hazards: weather and car accidents, alcohol and car accidents; eating, drinking, and car accidents

Hot and cold weather conditions

Biological hazards

Animals and insects

Bloodborne & waterborne pathogens

Personal (or Health) Hazards

Drug and alcohol abuse in the workplace

Ergonomic hazards

First aid and access to medical assistance

Lone worker

Respiratory hazards

Sanitation and disease prevention

Workplace Violence

PHYSICAL HAZARDS

Arc Flash

Description of hazard. Water plant operators and associated maintenance staff are frequently asked to work on electrical switchboards, panel boards, and motor control panels that are energized. In order to perform such work, an arc flash analysis must be done to determine the arc flash boundary and level of personal protective equipment (PPE) required to protect workers from potential exposure to an arc flash.

Common controls. Electrical safety awareness training for operators is usually included in most basic safety training and more advanced "qualified person" training must be provided for those who are required to work on energized systems. Based on new standards, additional training is needed on the following topics:

- Arc flash hazards (AFH) and how to avoid them
- Required PPE for AFH
- Basic arc flash analysis of the electrical system
- Understanding the arc flash warning label
- Risk hazard categories
- Safe work practices and troubleshooting procedures

Arc flash hazard analysis must be completed for the power supply network and equipment and equipment control panels must be field marked to warn of potential arc flash hazards; specific details on the flash boundary and appropriate PPE are required on the warning labels for qualified persons who may work on or near the hazard. An example of a warning sign that meets the ANSI Z535 Standard is in the electronic resources download. PPE for arc flash protection is required and depends on the incident energy at every point where a worker may work on energized equipment. PPE could include shoes, gloves, flame-resistant clothing, and safety glasses/shields. The selection of the proper PPE and training procedures depends on an accurate arc flash assessment being done on the electrical system. Professional services are available in most locations and should be used to ensure that all arc flash regulations are followed.

Regulatory information and other resources

OSHA regulations 1910.302–308 and 1910.331–335 require that all "non-dwelling" facilities have an arc flash hazard analysis done to determine the presence of a flash hazard, the arc flash boundary, and the level of PPE required.

The National Fire Protection Agency (NFPA) 70E regulations provide a roadmap to meeting the latest workplace arc flash protection requirements and meeting the OSHA standards.

Permit-Required Confined Space Entry

Description of hazard. Water plant operators and associated maintenance staff are exposed to possible death or injury when entering what is considered a confined space. Examples of confined spaces in the water industry are storage tanks, tunnels, vaults, meter pits, excavations, and process basins that are large enough for a worker to enter, have restricted means of entry or exit, and that may involve hazards related to atmospheric conditions, engulfment, configuration, or any other recognized serious safety hazard.

Common controls. Confined spaces must be identified and you must inform employees of the existence, location and hazards by inventory and labeling or by training and hazard evaluation before entry. Training for entrants and entry supervisors and atmospheric testing with direct-reading instruments are required before a permit required confined space is entered. Also, before an entry is allowed, a written entry permit must be completed and posted at the confined space location. In many situations, power ventilation equipment may be used to ensure adequate oxygen levels are maintained and toxic contaminants are not above permissible exposure levels. Figure 6-1 outlines a decision-making process for determining if a space is a permit-required confined space.

Rescue procedures must also be planned prior to the entry, and a minimum of one trained attendant must always remain outside the confined space to monitor the worker who has entered the confined space. Workers must be trained properly and equipped with the required safety equipment before the work is conducted.

Hazards from electrical equipment, mechanical equipment, chemical or pressurized piping and other energy sources must also be locked-out before working on it to ensure

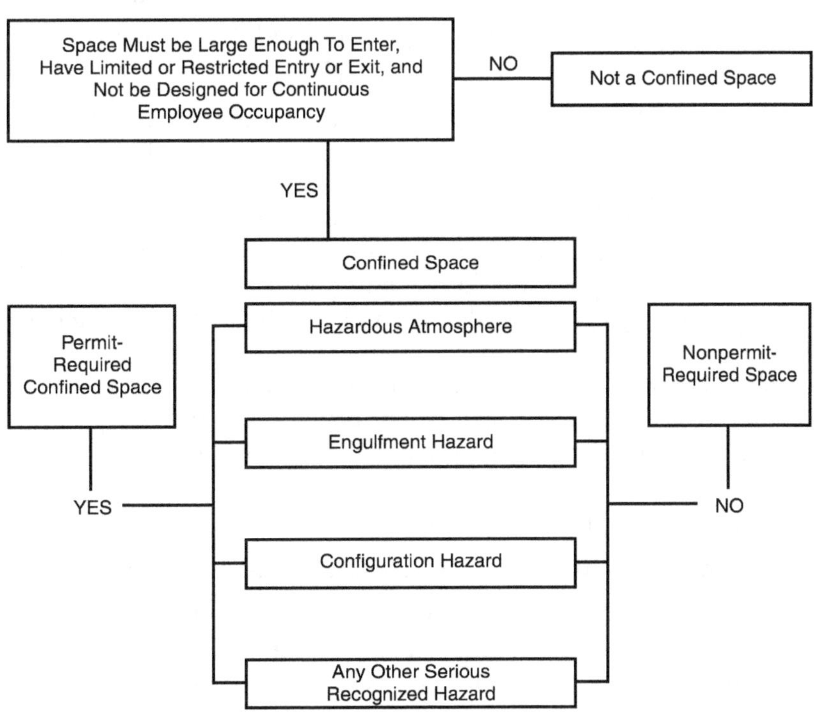

Courtesy of Melinda Raimann, Cleveland (Ohio) Division of Water

Figure 6-1 Confined space decision tree

worker safety. Valve pits or manholes that may be flooded with water or other fluids must also be rendered safe prior to entry. If work is performed on an elevated storage tank, additional fall protection procedures need to be followed. When workers are exposed to air contaminants such as sandblasting materials, welding fumes, or paint solvents, additional air monitoring and explosive hazards controls need to be taken.

Regulatory information and other resources

OSHA regulations (1910.146)

Excavations and Trenches

Description of hazard. Water distribution workers and associated inspection staff are frequently required to work in or around excavations and trenches. Hazards related to excavation and trenching operations include entrapment, suffocation, gas explosion, electrocution, and being struck by heavy equipment.

Common controls. A *competent person* must be trained to identify all existing and predictable hazards associated with the excavation or trench, including identification of soil types, and be authorized to take corrective action to eliminate hazards.

A written excavation safety plan needs to be developed to ensure compliance with existing safety standards. Components of the plan should include

- notification of the local "one-call" system for locating underground utilities;
- location of underground utilities;
- consideration of confined space atmosphere potential;
- selection of proper soil protective systems and personal protective equipment and clothing;
- determination of soil composition and classification;
- assessment of surface and subsurface water problems;
- determination of depth of excavation and duration of work; and
- training and supervision of all workers.

The competent person must be onsite and ensure the plan is followed. Daily and periodic inspection of the excavation and trench site must be conducted and documented. Conduct an inspection before every shift, rain event, or any other event that increases hazards. (A sample daily excavation safety checklist is in provided in the online resources.)

Soil testing must be done on freshly excavated samples from the work site; the testing determines the soil type and type of soil protective system that should be used. The three basic protective systems are sloping and benching, shoring, and shield (trench boxes) (Figure 6-2).

Sloping and benching systems for excavations 5- to 20-ft deep must be constructed under the instruction of the competent person. A registered professional engineer must design and stamp excavations greater than 20-ft deep to ensure safety. Other general hazard controls include

- store spoil piles a minimum of 4 ft from the side of excavation;
- remove or support all overhead hazards (e.g., utility lines);
- erect signs and barricades to protect workers from vehicular traffic;
- test for a hazardous atmosphere daily if the excavation or trench is 4 ft or deeper and a potential for such hazards exists;
- provide a means of egress from the excavation or trench with a ladder, ramp, or stairs so lateral travel distance does not exceed 25 ft;

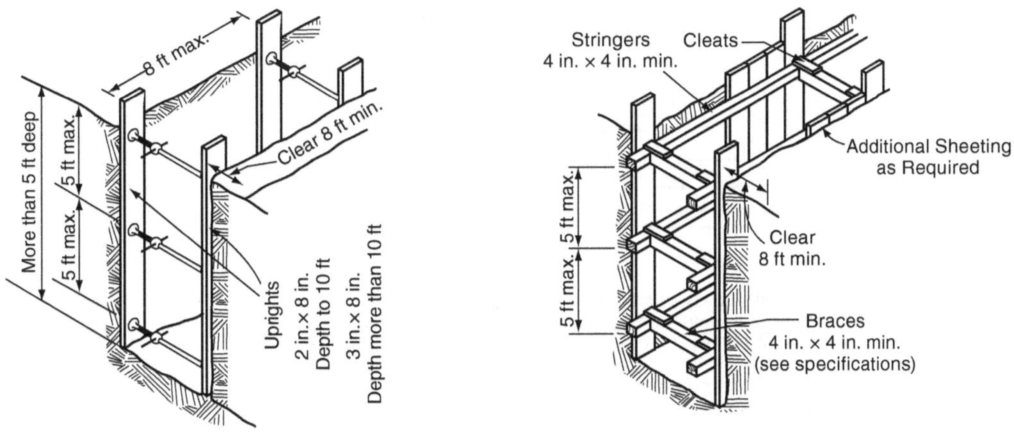

Source: AWWA Water Operator Field Guide

Figure 6-2 Hard, compact ground (left) can be braced at intervals, while saturated, filled, or unstable ground (right) needs additional sheeting or shielding to hold back loose soil.

- require workers to wear rescue harnesses with a lifeline attached; and
- alert rescue teams of the location of the excavation or trench when possible.

Regulatory information and other resources

OSHA regulation 1926.650 Safety and Health Regulations for Construction—Excavation
NIOSH training materials are available at www.cdc.gov.

Hazardous Energy Control (Electrical Safety)

Description of hazard. Water plant operators and associated maintenance staff are frequently exposed to many electrical hazards that can cause burns, arc blasts, fires, explosions, shocks, and electrocution (death). Overhead wires at various voltages are present on most water treatment facilities and throughout all service areas. The presence of water in the workplace compounds the hazard from electricity because water is a good electrical conductor. Electrical energy is the primary power source for all pumping and control equipment and good lockout/tagout programs are essential to avoid exposure to live circuits. Equipment maintenance and repair may expose staff to other hazardous energies, such as rotating equipment, high pressure, chemical contact, or falling heavy loads to name a few.

Common controls. General electrical safety practices that are essential to a safe working environment include

- electrical safety "qualified person" training and the use of electrical PPE for those who work on energized systems;
- using Lockout/Tagout programs to de-energize equipment before working on it, whenever possible;
- maintaining all electrical equipment according to prescribed practices and manufacturer's instructions;
- inspecting all electrical equipment on a regular basis and make repairs only if trained and authorized to do so, and reporting all electrical failures and repair needs to appropriate power suppliers;
- ensuring all safety features are used as designed;

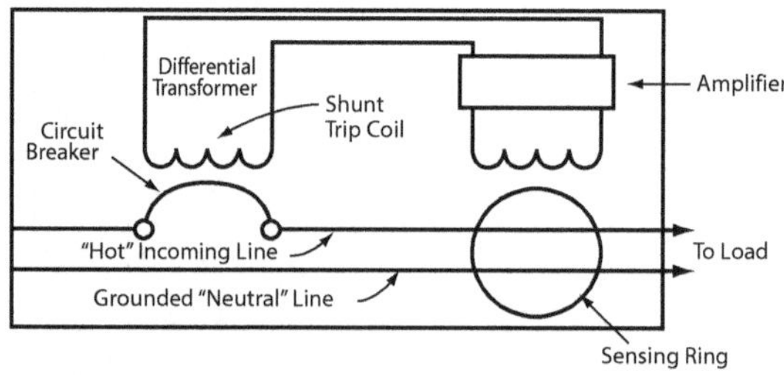

The differential transformer continuously measures the current flow in the "hot" and "neutral" lines. Under normal conditions, the current is equal in each line. If there is a difference of as little as 5 mA (0.005 A) the amplifier energizes the shunt trip coil which causes the circuit breaker to trip in 1/40th of a second or less.

EXAMPLE: A hand drill has a defective motor winding allowing a portion of the current to flow to the metal case and thus through your body causing a shock and possible electrocution.

Always use a ground fault interrupter when using electrical equipment outdoors and in damp, wet locations. Always make sure your electrical tools are in good shape.

Source: AWWA Water Operator Field Guide.

Figure 6-3 Ground fault interrupter

- protecting electrical cables and cords from damage;
- using extension cords only when necessary, but never as a substitute for permanent fixed wiring;
- avoiding wearing metal objects (rings, watches, etc.) when working with electrical systems;
- maintaining a minimum 10-ft (3-m) clearance from overhead wires that contain 50 kilovolts or less (Higher voltages require a greater clearance distance);
- when working in damp locations, inspecting electrical cords and equipment carefully to avoid shocks and always making use of the ground-fault circuit interrupter (Figure 6-3);
- avoiding use of electrical tools and lighting in wet and/or explosive environments, and using explosion-proof equipment and wiring instead in such conditions;
- ensuring that all electrical power tools are properly grounded or double insulated for protection from shock hazards;
- keeping doors on electrical panels in place and closed, correctly labeling circuits breakers and switches and plugging empty breaker slots; and
- using voltage-rated rubber gloves or jumpers when removing the water meter or repairing a water service within an excavation.

Electrical protective devices, including fuses, circuit breakers, and ground-fault circuit interrupters are critically important to electrical safety. Ensure all such devices are installed and maintained properly. Arc-blast (arc-flash) occurs when high-amperage currents jump from one conductor to another through air, generally when servicing equipment. See the arc flash hazard section for more information.

A Control of Hazardous Energy or "Lockout/Tagout" program is essential to controlling injury from both electrical energy and other things which are sometimes not recognized as hazardous energies. This can range from reopening a valve to a portion of the distribution system still under repair, not supporting a load when repairing a crane or other elevated equipment, opening a line with residual chemical or pressure, not chocking a fan or blower which can turn with a gust of wind, not blocking or compressing spring loaded equipment, not allowing hot equipment to cool down, not dissipating power stored in a capacitor, not isolating a large pipeline that can inundate a basin or a basement, etc. Control measures include

- a Lockout/Tagout (LOTO) program with energy-control procedures to ensure equipment is shut and energy is dissipated, the means of reactivating is locked/tagged, and employees have control over these to protect themselves when working;
- protective materials and hardware needed to lock and tag;
- provide LOTO authorized employee training on the safe application, usage, and removal of energy controls and "affected employee" training for those not authorized; and
- utilize written Energy Control Procedures for LOTO of equipment with multiple sources of hazardous energy requiring lockout, either prepared in advance of routine maintenance or developed as needed for unforeseen work (or by program intent).

Regulatory information and other resources

OSHA regulations 1910.301–1910.399 govern electrical work and 1910.332 covers training in safe work practices when working with electrical equipment.
OSHA regulations 1910.147 govern control of hazardous energy requirements for equipment with electrical, rotational and any other hazardous energy.
NIOSH training materials are available at www.cdc.gov/niosh.

Hot Work

Description of hazard. Hot work is one the largest causes of facility fires. Hot work is a temporary operation involving open flames or producing heat and/or sparks. The sparks, welding slag, open flames, and hot surfaces can act to ignite nearby combustibles, forming what could become a raging, uncontrollable fire. A fire in a hidden area can often go unnoticed until it is well established, spreading and causing needless destruction to nearby equipment and materials.

Common controls. A formal permit to authorize hot work is required for all cutting, welding, brazing, grinding, soldering, or any other activity that produces an ignition source outside of safe, designated welding/hot work areas. A written program should specify its use and the training needed for those given the authority to issue hot work permits. The permit should require that trained fire watchers maintain constant fire watch of the work area during the hot work and for one hour after work is completed, and provide for monitoring of the area for an additional three hours, when appropriate (e.g., where near wall or floor penetrations).

Additional hot work precautions include the following:

- As much as possible, remove all potential fire hazards from hot work area.
- Use guards to confine the heat, sparks, and slag, and to protect the immovable fire hazards.
- Welders and other hot work personnel must wear PPE, including eye protection, helmets and/or hand shields, and other protective clothing.
- Keep appropriately rated fire extinguishing equipment in the area, in a state of readiness for instant use.
- Vent all hollow spaces, cavities, or containers to allow air or gases to escape before preheating, cutting, or welding.

- Place welding cable and other equipment so that it is clear of passageways, ladders, and stairways.
- Never conduct hot work in the presence of explosive atmospheres (mixtures of flammable gases, vapors, or liquids or combustible dusts with air), or on a metal partition, wall, ceiling or roof with a combustible cover.

Confirm that cutters or welders and their supervisors are suitably trained in the safe operation of their equipment and the safe use of the process. Before issuing a permit authorizing hot work, consider whether the work can possibly be avoided, or if there is another, safer, way to complete the work.

A sample hot work permit is available in the online resources. The permit must be only valid for the single job and expire at the end of the shift or on completion of the fire watch.

Regulatory information and other resources

OSHA Regulation 1910.252, Welding, Cutting, and Brazing, addresses safety precautions for hot work.

Iowa State University Department of Environmental Health and Safety Hot Work Permit Program details a hot work program. www.ehs.iastate.edu/publications/manuals/hotwork.pdf.

NFPA 51B: Standard for Fire Prevention During Welding, Cutting, and Other Hot Work, provides guidance for people who manage, supervise, and perform hot work. www.nfpa.org.

Materials Handling

Description of hazard. Water plant operations and maintenance tasks include various materials handling activities. These work activities include the operation of cranes and hoists, powered industrial trucks, manual lifting, and other tasks related to water treatment and distribution. Workers can be at risk of collisions with moving parts and equipment, strain or injury from exertion or improper techniques, and other injuries.

Common controls. Workers must be properly trained to operate various types of materials-handling equipment. When cranes or hoists are used, a trained operator must be at the controls and all aspects of equipment inspection, maintenance, and operation procedures must be followed to ensure safety. Proper rigging for lifting is also required to prevent the load from shifting during the lift. When planning the lift, the operator must understand the characteristics of the load, including weight, dimensions, center of gravity, and the hazardous/toxic nature of material in the load.

Operators must also review the rigging to ensure slings and other devices are sufficient to support the load and attach it to the hook. During the lift, safe clearances must be maintained and workers must stay clear of the loads. Periodic inspections of all operating mechanisms, hydraulic lines, hooks, ropes, chains, and related electrical apparatus must be carefully documented. Any damaged or defective equipment must be immediately tagged and removed from service. All repairs and adjustments to this equipment must be performed by well-trained and certified workers.

Powered industrial trucks (forklifts) are common in many large water utility plants. All operators must be trained in the safe operation of the specific type of truck being used. Some basic rules for forklifts include the following:

- If the driver can not see past a load in front, then the forklift should travel backwards, carefully.
- Know the capacity and load limits for a vehicle, and never attempt to lift beyond them.

- Maintenance and repairs need to be conducted by authorized and trained personnel; call on trained mechanics when needed.
- When leaving a vehicle, lower the forks, put controls in neutral, set the brake, shut off the power, and remove ignition key or connector plug.
- Never park in front of fire equipment, doors, exits, or in high traffic areas.

Manual lifting techniques that avoid causing back injuries are essential (Figure 6-4). The following guidelines will help avoid back problems:

- Wear back braces if required to lift and carry heavy loads.
- Avoid twisting and turning movements.
- Use legs to position and move the torso.
- Avoid leaning or bending over for extended periods.
- Stretch and exercise the back before starting work each day.
- Inspect the work area for slip, trip, and fall hazards.
- Inspect steps and stairs before climbing; use a handrail whenever possible.
- Lift with the legs; position the body, so the load is centered and supported by the body before lifting and carrying.
- Avoid sudden jerks and pulls on a load, which could cause muscle sprain or injure discs.

Regulatory information and other resources

OSHA regulations 1910.176 –1910.184 govern materials handling.
NIOSH training materials are available at www.cdc.gov/niosh/topics.

Noise Exposure

Description of hazard. Water plant operators and associated maintenance staff are frequently exposed to noise levels that can be harmful to hearing. Noise sources include pumping and process equipment, alarms, generators, compressors, and ventilation equipment. Excessive noise exposure can cause temporary or permanent hearing loss, stress, and other physical problems.

Generally, noise levels above the 85-decibel level (dBA), expressed as an 8-hr time-weighted average, trigger the requirement for a hearing conservation program in the U.S.

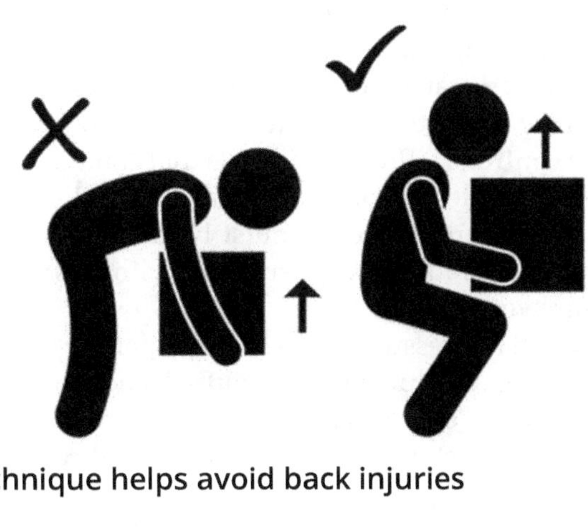

Figure 6-4 Proper lifting technique helps avoid back injuries

Certain impact or impulse noise levels may also be present and mandate the use of hearing protection.

Common controls. Noise can be controlled in the work environment by isolation or enclosure of the source of the noise. Control rooms can be constructed in treatment plants to provide a low-noise environment for operators. If operators must enter high-noise areas they can be required to use personal hearing protection to avoid hearing loss.

Hearing conservation programs are mandatory when worker's noise exposure levels equal or exceed a prescribed regulatory dose (time-weighted average based on 8-hr time interval). A hearing conservation program includes noise level monitoring, initial and annual audiometric testing of workers, mandatory use of hearing protection that is properly fitted and maintained, and annual training on the harmful effect of noise. Hearing protection must be made available to workers who are exposed above the action level and required when the levels of noise exceed standards, generally a time-weighted average of 90 dBA. OSHA calls for subtracting 7 dBA from the Noise Reduction Rating (NRR) of any ear plugs and earmuffs commonly used for hearing protection and others recommend derating the result further by a factor of two to account for variations in users. Depending on the method used and devices, noise exposure levels can usually be reduced by 10–20 decibels.

Regulatory information and other resources

OSHA regulations 1910.95
NIOSH training materials are available at www.cdc.gov/niosh/topics.

Pressure Vessels

Description of hazard. A pressure vessel is a cylindrical or spherical metal container capable of withstanding pressures exerted by the contained material. Many liquids and gases must be stored under high pressure in such vessels. Safety codes have been developed that specify the container design for specified conditions, with special emphasis on the strength of the vessel to prevent rupture and subsequent explosions. All shells, test chambers, tanks, and model parts designed for internal pressures greater than 15 psi are considered pressure vessels.

Pressure vessels include compressed gas storage tanks (e.g., air, oxygen, nitrogen tanks), anhydrous ammonia tanks, autoclaves, chemical reactors, hot water storage tanks, hydro pneumatic tanks, and refrigerant vessels, designed for a pressure greater than 15 psi and a volume greater than 5 cubic feet in volume or one and one-half cubic feet in volume with a pressure greater than 600 psi.

When pressure vessels fail, the result is usually catastrophic: a ruptured vessel releases vast stores of energy and contents. The results may produce a shock wave of the vessel contents, and high-speed projectiles. Such failure poses a safety risk to personnel, jeopardizes other critical equipment and material, and may result in the release of hazardous or toxic contents, environmental pollution, and loss of production.

Common controls. Pressure vessels that contain only low pressures are constructed of tubes and sheets rolled to form cylinders. The walls of pressure vessels whose contents are under high pressure, however, must be thick enough to provide adequate strength to avoid rupture. Hydraulic and pneumatic cylinders are machined components that are types of pressure vessels.

To ensure that pressure vessels are operating and maintained safely, periodic inspections must be conducted by qualified and authorized personnel at a frequency depending on location (e.g., every two years). Internal inspections are conducted if deemed necessary by the inspector of record. There are three types of board-certified pressure vessel inspectors:

- Boiler and Pressure Vessel Inspector: an employee of the state or local boiler safety regulatory agency who is authorized to inspect any boiler or pressure vessel subject to the ASME Boiler and Pressure Vessel Code (BPVC)
- Special Inspector: an insurance company employee who is only authorized to inspect what the inspector's company insures
- Owner–User Inspector: an employee of the organization that operates the boilers or pressure vessels who is trained under the company's authorized inspection program and supervised by one or more engineers with qualifications satisfactory to and approved by the boiler safety agency having jurisdiction for that facility.

To be authorized to perform repairs, a repair company must hold an "R" stamp issued by the National Board of Boiler and Pressure Vessel Inspectors. In all cases, the owner and the repair company must consult an inspector before a repair is begun. The completed repair is subject to the inspector's acceptance.

Pressure vessel relief valve inspection and testing (or replacement) frequency is generally determined by the National Board Inspection Code and is typically five years or less.

If an accident does occur, a detailed report of the accident must be submitted immediately to the regulatory agency. If someone is injured or the tank explodes, neither the vessel nor any parts shall be removed or disturbed before the chief inspector grants permission, except for the purpose of saving human life and limiting consequential damage.

Regulatory information and other resources

Pressure vessels must be designed in compliance with the latest edition of the American Society of Mechanical Engineers (ASME) Boiler and Pressure Vessel Code (BPVC), Section VIII or Section III.

Pressure-vessel welding must be in compliance with the BPVC Section IX for welding qualifications and Section V for welding inspection.

OSHA regulations 1910.101 through 1910.111 addresses pressure vessels used for compressed gases.

The National Board of Boiler and Pressure Vessel Inspectors has a synopsis of state and city license and inspection requirements (https://www.nationalboard.org/ViewAllSynopses.aspx). The National Board Inspection Code is its standard for inspection and testing, including relief valves.

Tractor Operations

Description of hazard. Tractors are versatile tools that can be used for many different tasks. Unfortunately, when used improperly, tractors can and do cause many severe injuries and fatalities. Many accidents are similar, such as rollovers, flip overs, and back overs. Most tractor incidents are caused directly or indirectly by operator error—bad judgment, unnecessary haste, or carelessness. In many cases, tractor operators work independently without much supervision, so operators must understand and buy into the importance of safety. To promote this safety culture, management must stress safety on a regular basis.

Common hazards and accidents in tractor operation of are

- rollovers;
- flip overs;
- caught in/between pinch points;
- power take off (PTO) hazards;

- fueling;
- back overs/struck by;
- front-end loader incidents;
- incidents with attachments;
- falls from the equipment;
- hydraulics;
- shear and cutting points;
- thrown objects;
- highway incidents;
- hot surfaces; and
- noise.

Tractor rollovers happen when the center of gravity moves past a baseline of stability, either to the side or rear of the machine. Many rollovers happen at speeds less than 8 mph and on slopes less than 5 degrees. A tractor can flip in less than one second, so once it reaches the point of no return, it is too late.

Contributing factors to rollovers include the various dimensions and weights of implements, operating speed, type of terrain, and moisture in soil.

Common causes for rearward rollovers:

- Tractor is stuck in mud or soft soil, which prevents the rear wheels from rotating
- Climbing a hill that is too steep
- Releasing the clutch too quickly with the transmission in a low gear and engine at a high speed

Side rollovers can be caused by;

- driving on a hillside that is too steep;
- driving too close to the edge of a roadside ditch or embankment; and
- a front-end loader that is elevated too high on a hillside or in a turn taken at excessive speed.

Common controls. Tractor safety has improved greatly over the last few years. Rollover protective structures, seatbelts, improved guarding, enclosed cabs, electronic controls, stability sensors, and climate control have all contributed to making tractor operations safer and more comfortable.

Prior to operating a tractor, employees should assess the work area for obstacles, holes, slopes, ditches, curbs, wire, glass, and bees. Conduct tractor pre-operation checks, and notify a supervisor if defects are found. Employees should wear PPE and always maintain and use the tractor's safety features (e.g., rollover protective structures [ROPS], seatbelts, and PTO shields). Operate the tractor and implements safely and according to manufacturer's instructions. Although tractors are versatile machines, other machines may be better suited to specific tasks and should be used instead. Know the equipment's blind spots, and never carry passengers.

ROPS work by limiting a rollover to 90 degrees and preventing the operator from being crushed under the weight of the tractor. ROPS work only if a seatbelt keeps the operator in the cab as the tractor is rolling.

Proper training should include classroom and practical lessons and should be documented in writing. This training should include

- tractor and implement hazards;
- review of operator's manuals;

- the importance of not operating defective equipment;
- location of guards, shields, and other safety devices;
- identification of safety hazards at the work site;
- hands-on supervised instruction;
- competency test on equipment and implements; and
- instruction on attaching implements.

Tractors generate deadly carbon monoxide, so they should not be used in buildings. If operation in buildings is required, be sure there is enough ventilation to eliminate the possibility of carbon monoxide buildup.

Safety equipment that should be on all tractors:

- Seatbelt, if equipped with roll-over protection (ROP)
- Rearview mirrors
- Head/tail/turn lights
- Back-up alarm
- Slow vehicle warning triangle
- Fire extinguisher (5 lb multipurpose – ABC class)
- High-intensity warning lights

Prevention of side rollovers:

- Slow down when cornering (centrifugal force pivots a tractor on the outside wheels).
- Avoid working on steep slopes if possible.
- Stay at least one tractor width away from the edge of a bank.
- Follow operator manual instructions on operating equipment on slopes.
- Make wide slow turns.
- Know the terrain: small holes, depressions, large rocks, or stumps can cause tip over.
- Go down slopes in low gear.
- Back up a slope if transporting a heavy implement.
- Follow the tractors operator's manual and instructions on side mount implement placement (up or down hill).

Prevention of flip overs:

- Release the clutch only when rear wheels can rotate forward.
- Do not climb steep hills in a forward direction. Instead, back the tractor up the hill.
- Use only enough engine speed to start tractor moving while engaging the clutch smoothly.
- Change speed gradually by applying power smoothly.
- Ballast (counter-balance) the tractor properly for the job.
- Hitch loads properly to a drawbar, do not attempt to pull logs or other objects that may catch on an obstruction in the ground and pull the front end off the ground. Use the drawbar at the height recommended in the operator's manual. Do not alter or raise the height of the drawbar.
- Use reverse gear to break tractor tires free from frozen conditions.
- Use another machine to pull a tractor out of the mud.

When traveling on the road in a tractor:

- Attach a slow-moving vehicle emblem.
- Use headlights and high-intensity flashing lights.
- Use an escort vehicle when feasible.
- Secure attachments in the transport position. Never operate attachments in transit.
- Take it slow and pull over when needed to let normal traffic pass.
- Take special care at intersections, turns, and curves.
- Always leave room to stop safely.
- Carry a cell phone.

Regulatory information and other resources

OSHA regulations 1926.1001 and 1926.1002 describe the requirements for installation of ROPS on heavy equipment used in construction.

Recommendations Specific to Hazards Associated with Heavy Equipment and Powered Industrial Truck Use. www.osha.gov/SLTC/etools/hurricane/heavy-equip.html

Traffic and the Work Zone

Description of hazard. According to the Bureau of Labor Statistics, transportation incidents and workers struck by vehicles or mobile equipment account for the highest number of fatal work injuries. In a work zone, workers are susceptible to the careless driving of oncoming traffic as well as to the actions of work vehicles and equipment.

Common controls. Effective temporary traffic controls must provide for the safety of workers, road users, and pedestrians, while providing for the efficient completion of the work activity (Figure 6-5). Construction area warning signs must be visible at all times when work is being performed and removed or covered promptly when the work is complete. Hazards associated with working on the roadway cannot be completely eliminated because of the constant shifting and changing nature of work zone activity, but they can be greatly reduced by implementing the following controls:

- Train all employees on work zone traffic safety.
- Evaluate the work space in advance to ensure work area protection and minimize possible disruptions to pedestrians and vehicular traffic.
- When exposed to vehicular traffic, employees should wear high-visibility safety apparel.
- Design and implement a traffic control plan (TCP) according to the characteristics of each location, such as speed, volume, direction of traffic, time of day, type of work, and duration of work.
- Set up advance warning signs to alert oncoming drivers of the work area. Signs must convey a simple, easily understood message.
- Use channeling devices (cones, barricades, etc.) in the transition zone for smoothing vehicular traffic flow and to separate vehicular traffic from the work space.
- Monitor traffic flow on a continuous basis and be aware of changing conditions of the work area or traffic and make adjustments accordingly.
- When feasible, place reserve trucks and other work vehicles in the work zone in a manner to provide additional protection between oncoming traffic and the work area.

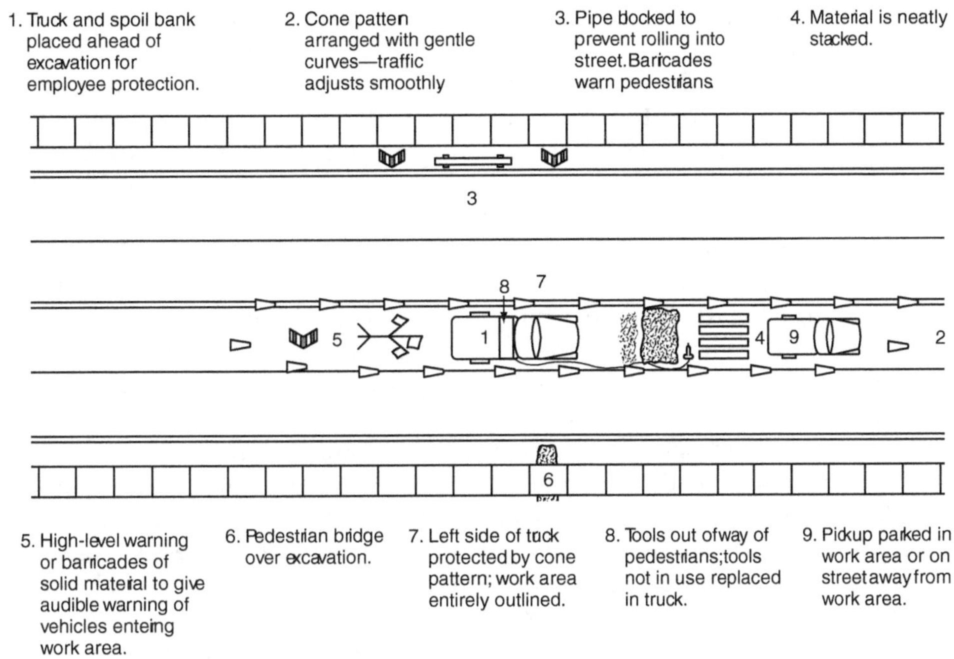

Source: AWWA Water Operator Field Guide

Figure 6-5 Proper placement of cones, vehicles, pipes, and equipment help keep a field operation safe for workers, pedestrians, and public traffic

- Keep equipment and material inside the work zone to prevent employees from entering an unprotected area adjacent to the work zone to access the equipment.
- Promote awareness of the dangers of working around vehicles and moving equipment in both the work area and outside the protected activity zone.

In addition to a TCP for outside of the work zone, an *internal* traffic control plan (ITCP) should be created for the work zone itself. ITCPs are used to safely coordinate and control the interaction between workers, construction vehicles and operating equipment within the work space where more than a single work group or activity is required to complete the work.

Regulatory and other resources

OSHA regulations 1926 Subpart G—Signs, Signals, and Barricades

Federal Highway Administration's *Manual on Uniform Traffic Control Devices for Street and Highways,* which dictates traffic control signs or devices used for protection of construction workers and sets minimum standards, provides guidance, and ensures uniformity of traffic control devices to promote highway safety and efficiency

ANSI/ISEA 107-2004 'American National Standard for High-Visibility Safety Apparel and Headwear' Class 2 or 3 requirements

Driving Hazards

Description of hazard. *Alcohol and car accidents.* Almost 11,000 people die every year in accidents related to drunk driving. Of those, more than three youths die every day. The average for a drunk driving–related death is one person every 50 minutes. These deaths make up about one third of all traffic deaths. In terms of distance, a drunken person needs an estimated four extra feet of braking distance compared to a driver who has not had anything to drink.

Cell phones. Cellphones have changed the world in so many ways. They connect people across the county and even the world. They allow immediate contact with emergency services, no matter where an accident occurs, which can save lives. Unfortunately, using cellphones while driving is a main reason why people need to call for help in the first place. According to autoinsurance.org, cellphones are responsible for 24,000 injuries per year and 995 deaths.

Twenty-one percent of drivers admit to texting while driving and a 46 percent of teens admit to texting while driving. Texting while driving increase chances of crashing by a factor of eight, which is twice as bad as getting behind the wheel while drunk. A texting driver need 70 extra feet of extra braking distance to stop a car, compared to an attentive driver.

Weather and car accidents. The weather is also a powerful foe. Icy or wet weather creates slippery, unsafe roads and the risk of hydroplaning. Snow-packed and slushy roads are difficult to navigate. Precipitation, fog, and dust storms impede vision. Wind can push light or high-profile vehicles off course. Add speed that is excessive for the conditions and an accident is bound to happen. More than 1.5 million weather-related accidents occur every year in the United States, resulting in 673,000 injuries, with 7,400 deaths.

Eating, drinking, and car accidents. Splitting attention between the road and food or beverage selection can also be dangerous. Studies show that eating or drinking can increase the chance of an accident by 80 percent, and 70 percent of people admit to eating while driving and 80 percent acknowledge drinking while driving.

Backing accidents and parking lots. Backing with a large vehicle with limited rear visibility is always difficult and there is a higher risk of accidents in busy locations and parking lots. An empty parking lot when you arrive is no guarantee that you will not be "boxed in" later.

Common controls. No one can control another driver. Alertness and attention to the road, other drivers, and surrounding conditions are key to safe navigation of the roads and avoiding the hazardous behaviors of other drivers and the weather. Avoid texting, talking on the phone, and eating or drinking while driving. Do not consume alcohol or drugs then get behind the wheel. Alter driving habits according to the weather and leave extra time to arrive at a destination. Park in emptier areas of parking lots and in a pull through location or back in to allow pulling forward to exit whenever possible. Turn lights on during hazardous weather and at dawn and dusk.

Employ these defensive driving techniques:

- Scan the roadway and adapt to conditions.
- Anticipate the actions of other drivers.
- Use a turn signal so other drivers know what to expect.
- Know a vehicle's stopping distance and use the two-second rule for following distances. (Remain at least 2 seconds per 5 mph from the vehicle in front to provide a distance of one car length stopping distance.)
- Respect other users of the roadway, and share the road with all travelers, including bicyclists.
- Pass only when allowed and there is no oncoming traffic.
- Yield the right of way.
- Be alert and avoid distractions.

Fleet purchases should always include backing alarms and should consider adding backup cameras and hands free speaker phone integration.

Regulatory information and other resources

National Highway Traffic Safety Administration: www.nhtsa.gov/Driving+Safety

Official US government website for distracted driving: distraction.gov

Insurance guidelines for safe driving: auto insurance/driving-hazards

Hot and Cold Weather Conditions

Description of hazards. Working in extreme weather or conditions—hot or cold—can be dangerous. People who work outdoors must be aware of the dangers, the signs of injury, and the way weather hazards can be controlled.

Hot weather hazards

High temperatures + High humidity + Physical work = Heat illness.

When the body is unable to cool itself through sweating, serious heat illnesses may occur. Operations involving high air temperatures, radiant heat sources, high humidity, direct physical contact with hot objects, or strenuous physical activities have a high potential for inducing heat stress in employees. The most severe heat-induced illnesses are heat exhaustion and heatstroke.

Heat exhaustion should not be dismissed lightly. Fainting from heat exhaustion can be dangerous, not only because the victim is obviously ill, but also because the victim may be operating machinery or controlling an operation that should not be left unattended or that could pose an additional danger to the victim. The symptoms of heat exhaustion include headache, nausea, vertigo, weakness, and thirst.

Heatstroke occurs when the body temperature rises to critical levels, causing a medical emergency. If the body temperature is too high, death can occur. The symptoms of heatstroke include dry pale skin with no sweating, hot red skin that appears sunburned, rapid heartbeat, shallow breathing, mood changes such as irritability, and confusion.

Common controls. A victim of heat illness should be moved to a shady or air-conditioned area and their outer clothing removed. If heatstroke is suspected, call 911. Wet the person's skin with cool water by applying damp towels or spraying the victim. If the victim is able, have him/her drink cool water or other nonalcoholic beverage without caffeine.

Do not allow the person to leave the site or be left unattended unless authorized by a medical professional. The medical outcome of an episode of heatstroke depends on the victim's fitness and the timing and effectiveness of first aid treatment.

Controlling heat on the job is the first step in preventing overheating. Common controls include

- ventilation, air cooling, fans shielding, and insulation in hot work environments;
- frequent rest breaks in a cooler environment;
- drinking small amounts of cool (50–60°F/10–15.5°C) water or nonalcoholic and non-caffeinated liquid frequently—one cup every 20 minutes;
- additional breaks for anyone showing signs of heat-related strain or fatigue, particularly new employees or workers just back from vacation or illness who are not yet used to the heat environment;
- doing heavy work during the coolest part of the day;
- time to acclimate workers to heavy, hot work;
- dedicated first-aid training on heat disorders in every hot work area or for every work crew; and

- a buddy system for workers in chemical protective clothing to observe each other for any medical problem.

Cold weather hazards. When the body is unable to warm itself, serious cold-related illnesses and injuries may occur, including permanent tissue damage and possibly death. Prolonged exposure to freezing or cold temperatures or cold-water immersion may cause trench foot, frostbite, and hypothermia.

Hypothermia occurs when the core body temperature drops below 95°F (35°C). Symptoms of hypothermia include uncontrolled shivering, slurred speech, clumsy movements, fatigue, confused behavior, weak pulse, and slow, shallow breathing. If any of these signs are observed, call for emergency help.

Frostbite occurs when deep layers of skin freeze, becoming hard and numb, and usually affects the extremities—fingers, hands, toes, feet, ears, and nose. Symptoms of frostbite include hard, pale, and cold skin; pins and needles feeling, followed by numbness; and aching or throbbing of the affected area.

Trench foot occurs when the feet become swollen and itchy, then very painful, and leg cramps may occur. The skin may blister and die. It may involve the tip of the nose and ear as well. Workers are at increased risk of cold exposure when they have certain health conditions, such as cardiovascular disease, diabetes, and hypertension; they take certain medications; they are in poor physical condition; or they are older.

Common controls. Seek medical attention immediately for hypothermia, and quickly for cases of frostbite and trench foot. If medical help is not immediately available for hypothermia,

- Move the victim to a warm, dry location if possible; if not, shield him or her from the cold and wind as much as possible, including insulating the body from cold ground.
- Limit movements to only those that are necessary. Do not massage or rub the person. Vigorous or jarring movements may trigger cardiac arrest or further damage affected skin.
- Remove wet clothing; cut it away if necessary to avoid excessive movement.
- Cover the victim with layers of dry blankets or coats, including the head, leaving only the face exposed.
- Share body heat by lying unclothed under the covers next to the victim, making skin-to-skin contact.
- Monitor breathing. If the victim's breathing has stopped or appears dangerously low or shallow, begin cardiopulmonary resuscitation (CPR) immediately.
- Provide warm beverages if the affected person is alert and able to swallow.
- Apply warm compresses only to the victim's neck, chest wall, or groin but *not* to the arms or legs, which forces cold blood back toward the heart, lungs and brain, causing the core body temperature to drop.

First aid for frostbite or trench foot includes the following:

- Soak the affected area in warm water (102 to 110°F [39 to 43°C]) or repeatedly apply warm cloths to affected ears, nose, or cheeks.
- Wrap the frostbitten areas with dry, sterile dressings; wrap between frostbitten fingers or toes to keep them separated.

If protection from refreezing cannot be guaranteed, delay the initial rewarming process until a warm, safe location is reached to prevent thawing and refreezing, which may cause more damage.

Controlling cold conditions on the job is the first step in preventing frostbite and other cold-induced conditions. Ways to keep warm and safe include these:

- Wear proper clothing, particularly insulated, layered clothing, gloves, and hats.
- Train workers about cold-induced injuries and illnesses and appropriate first-aid techniques.
- Take frequent short breaks in warm dry shelters.
- Avoid exhaustion or fatigue; energy is needed to keep muscles warm.
- Use the buddy system to recognize danger signs, especially when the temperature is 10°F (–12°C) or below and when working near water.
- Drink warm beverages but avoid caffeine and alcohol.
- Eat warm, high-calorie foods.
- Provide heated trailers, shelters, or other warm areas for workers.
- Schedule outdoor work during the warmest part of the day, and plan a work/rest schedule that avoids prolonged exposure to the cold.
- Take more breaks when the temperature is low or the wind speed is high.
- Shield the work area from wind and draft.
- Allow workers to become acclimated to the cold before working full time in a cold environment.
- Keep an extra supply of clothing in the vehicle for workers who must travel for extended periods in the cold weather.
- Replace fluids regularly, particularly with warm liquids, because working in cold dry air can cause more water loss than usual.
- Review cold and heat stress policies annually.

Other resources

Mayo Clinic, Patient Care & Health Info, Diseases and Conditions. www.mayoclinic.org/diseases-conditions

National Institutes of Health, U.S. National Library of Medicine. www.nlm.nih.gov/medlineplus/ency

BIOLOGICAL HAZARDS

Animals and Insects

Description of hazard. Water plant operators and associated maintenance staff are frequently exposed to vector-borne diseases from the bites of infected ticks and mosquitos. One of the most common tick-borne diseases in the United States is Lyme Disease. Ticks are found in wooded areas, high grass, or leaf matter. West Nile Virus, which affects about one in five people who are infected, is carried by mosquitos that can be found near standing water or in weedy or wooded areas.

Rodents and other wild animals can also carry disease. Both live and dead animals can carry rabies and Rat Bite Fever. Hantavirus has also been found in rodent droppings, and infection can lead to Hantavirus Pulmonary Syndrome (HPS), which can be fatal. Exposures to animal urine, particularly from rodents, livestock and wild animals may carry *Leptospira* bacteria where human contact via contaminated water may result in Leptospirosis, a sometimes-fatal infection that is generally difficult to diagnose clinically.

Symptoms of vector-borne diseases often occur one to five weeks after exposure and can include fever, headaches, fatigue, body and muscle aches, vomiting and diarrhea, joint pain, rash, stiff neck and, paralysis.

Stinging insects that fly (e.g., bees, wasps, and hornets) can cause reactions that range from mild discomfort to severe allergic reactions requiring immediate first aid or medical care. Fire ants can bite and sting, injecting venom that causes a burning sensation and red bumps at the sting site, which develop into white fluid-filled pustules in one to two days. Poisonous snakes and spiders, such as the black widow and brown recluse, are frequently encountered by meter readers and installers and other distribution system workers.

Common controls. Prevention includes good housekeeping and grading in facilities and around distribution system access points. Both organic debris and man-made material accumulations (e.g., drums stored outdoors) can harbor vectors and hold sufficient shallow pools of rain water that will remain wet for the three days needed to hatch mosquito eggs.

If a worker is bitten or stung by an insect or rodent and symptoms of severe illness are apparent, or bitten by a poisonous snake, seek immediate medical attention. Otherwise, monitor the victim over time to determine if symptoms of illness occur.

Numerous steps can be taken to avoid vector-borne diseases and poisonous critters.

- Wear light-colored clothing, hat, long-sleeved shirt, and long pants tucked into boots to cover as much of the body as possible.
- Use insect repellents containing 20 to 50 percent DEET or Picaridin on exposed skin and clothing.
- Check skin and clothing for ticks daily, especially hair, underarms, and groin.
- Avoid working outdoors during dawn and dusk hours when mosquitos are usually most active.
- Do not disturb ant mounds and check over an area for ants before starting to work.
- Workers with a history of severe allergic reactions to insect bites or stings should carry an epi-pen (auto-injector) and advise co-workers of this condition.
- Avoid direct contact with rodents and wear protective clothing—including respiratory protection and gloves—when removing dead animal remains or droppings.
- Wear boots that are at least 10-in. high, and watch for snakes sunning on fallen trees, limbs or other debris.
- Use flashlights to scan work area for spiders before entering meter pits or vaults and wear long-sleeved shirt and long pants.

Other resources

Local college or university extension programs typically have information about pest control specific to an area and the prevalence of specific vector-borne diseases.

Centers for Disease Control and Prevention. www.cdc.gov

Waterborne Pathogens (also see Sanitation and Disease Prevention)

Description of hazard. Waterborne pathogens of primary concern come from direct contact with or aerosols of sewage/biosolids and include human enteric viruses, bacteria, parasitic protozoa and helminth eggs (to intestinal worms, flukes and nematodes). Pathogens from animal excreta (zoonotic pathogens) also include enteric bacteria (e.g., pathogenic *E. coli*, *Campylobacter* and *Salmonella* species) and parasitic protozoa (e.g., *Cryptosporidium* and *Giardia* spp. and some helminths) and should be assumed to be present where animal manure is generated and downstream in waterways. Many drinking water plants in the United States and Canada utilize surface water impacted by runoff from animal manures and wastewater/stormwater discharges, so assume waterborne pathogens may be present in raw waters and backwash waters/solids. Human-infectious enteric viruses (such as *Norovirus,* human adenoviruses, hepatitis A & E viruses, and various enteroviruses) are the pathogens present at the highest concentrations in sewage and typically cause infections at the lowest doses (10–100 viruses ingested). While most enteric pathogens result in mild gastrointestinal diseases, some also cause severe to life-threatening illnesses via liver, kidney, heart, blood and central nervous system infections.

Common controls. Basic hygiene steps like access to toilets, washing facilities and lavatories and washing hands with soap and water before eating or leaving the workplace is an essential control for sewage workers that also applies for water system employees who may be similarly exposed. If a worker is splashed or aerosolized exposed to aerosolizes of aerosolized sewage-impacted water, immediate clean water wash affected skin and if soaked through, washdown and change external clothes. A change of clothes before leaving the workplace is a prudent precaution for reducing even non-soaking exposures to reducing the spread of waterborne pathogens to those at home. Protective gloves, goggles, and splash suit should be used for tasks expected to result in contact with skin, eyes, and clothes. Where aerosol formation is expected (e.g., moderate to high pressure water spray) other PPE such as a dust/aerosol respirator or loose fitting facepiece, face shield, and impervious coveralls should be considered. This PPE must be made available for use in the field and at the work place if these cleaning/maintenance. hazards are anticipated. Disposable gloves should be used for sampling of any sewage-contaminated samples and hand disinfectants available at appropriate locations in the work environment.

Other resources

An excellent on-line compendium of information on waterborne pathogens can be found at the Global Water Pathogen Project website (http://www.waterpathogens.org); as well as specific information from the Centers for Disease Control and Prevention. www.cdc.gov

Bloodborne Pathogens

Description of hazard. Bloodborne pathogens are infectious microorganisms present in blood that can cause disease in humans. Workers who are exposed to these pathogens are at risk for hepatitis B virus, hepatitis C virus, human immunodeficiency virus (HIV), and other diseases. Water system workers trained in and required to perform first aid who can reasonably be anticipated to come into contact with blood as a result of doing their job duties are covered by the OSHA Bloodborne Pathogens Standard. Laboratory staff may also become exposed when working with sharp implements or beakers that may break and cut a worker. At small work places, employers often opt to not require any of its employees to perform such work and rely solely on outside occupational health and emergency medical providers. In that case, awareness training and post-exposure actions are necessary due to both the potential for an incident with blood release and exposure due to "Good Samaritan" acts (e.g., assisting an employee who has a nose bleed or cut).

Common controls. If occupational exposure to potentially infectious materials can be reasonably anticipated, employers must identify either job positions or positions and tasks for which contact with bloodborne pathogens may occur. An Exposure Control Plan is required to identify the jobs/tasks and hazards and identifies engineering controls (e.g., properly labeled disposal containers for sharps objects and infectious waste) and administrative controls (e.g., PPE).

First aid kits should include appropriate PPE, such as gloves and eye protection, to prevent exposure to blood and other potentially infectious materials.

Other steps to protect against worker infection by bloodborne pathogens include

- establishing a written exposure control plan and updating it annually;
- making vaccinations for hepatitis B available to all workers with occupational exposure;
- making post-exposure evaluation and follow-up available to exposed workers who experience an exposure incident; and
- providing information and training to workers and maintaining worker medical and training records.

Regulatory information and other resources

OSHA regulation 1910.1030

OSHA fact sheets and training information are available at www.osha.gov.

CHEMICAL HAZARDS

Chemical Exposure and Highly Hazardous Chemicals

Description of hazard. Water plant operators and associated maintenance staff are often required to work with chlorine, chlorine dioxide, ozone, ammonia, and other highly hazardous chemicals that are toxic by inhalation and can migrate to cause injury at a distance. Accidental releases of these chemicals have been reported by water treatment facilities across the United States and Canada. Such releases pose a significant threat to water operators and emergency responders and can even cause harm in the surrounding community. Acids, bases, and many other chemicals present significant risk of contact/splash burns and localized inhalation hazards due to off-gassing from liquid releases. "Inherently safer chemicals" like sodium hypochlorite ("Hypo"), aqueous ammonia, hydrofluorosilicic acid, and others are both significant contact hazards and have often been involved in incompatible mixing events that result in generation of toxic chemical off-gases. Many toxic gas exposure incidents at water plants attributed to chlorine gas are actually due to accidental mixing of Hypo and acidic chemicals (e.g., hydrofluorosilicic acid, ferric chloride, sulfuric acid). For example, a 2015 incident in North Canton, Ohio that sent four to the hospital was due to mixing Hypo and hydrofluorosilicic acid but several news articles identified the chemical only as "chlorine." Incidents involving a gallon or less have caused injury when working in an enclosed space such as when using hydrochloric acid for injector scale removal when residual Hypo is present. One 2019 incident at a Burlington, Mass., restaurant caused by mixing Hypo with scale remover from gallon containers for "better cleaning" resulted in one death and 13 treated. Where hospital and industrial wastewaters are potentially present, radioisotopes used for diagnostic and therapeutic procedures are of potential, but generally very low concern. These include technetium-99m (Tc-99m), Iodine-131 (I-131), Iodine-125 (I-125), Iodine-123 (I-123), Flourine-18 (F-18), Tritium (H-3), and Carbon-14 (C-14), and while uses at facilities need to have sufficient infrastructural and manpower resources to keep ambient radiation levels within specified safe limits, accidents may increase levels in facility wastewater discharges.

Common controls. To prevent against unexpected highly hazardous chemical releases, water suppliers must establish a comprehensive Process Safety Management (PSM) program and/or Risk Management Program (RMP). Key requirements include process hazard analysis (a careful review of what could go wrong and what safeguards must be implemented to prevent a release of hazardous chemicals), chemical/equipment design information, operating and maintenance procedures, training, and emergency planning.

"General duty" under various laws and OSHA regulations require many of the same measures for all chemicals including operating procedures, training and written hazard identification and control materials made available to those who work with these chemicals. The most critical protection for "inherently safer" chemicals is training and procedures to prevent accidentally unloading the wrong chemical into a container or tank (e.g., by locking and signs at fill ports, requiring operator presence for truck unloading) or mixing any two chemicals in an unauthorized manner. Water systems must complete a personal protective equipment (PPE) hazard assessment identifying PPE required for tasks involving chemical hazards (and all other hazards as well) and develop and implement a formal hazard communication program. PPE may include safety glasses for routine work, augmented by gloves, face shield, and chemical resistant suit and boots as needed for higher hazard tasks like chemical transfers and opening equipment for maintenance. The use of PPE specified in the assessment must be communicated and provided to employees, when required. Hazard communication includes requirements for labeling, and providing safety data sheets (SDS) and training for each water system worker. An example SDS and description of the required information that must be provided is among the electronic resources for this manual, available at www.awwa.org/M3. Also see "Respiratory Protection."

Regulatory information and other resources

OSHA regulation 1910.1200 for Hazard Communication and 1910.132 for Personal Protective Equipment

OSHA regulation 1910.119 for Process Safety Management and Appendix A (list of highly hazardous chemicals) and Appendix C (compliance guidelines and recommendations that can be used to meet the PSM Standard requirements)

PSM Covered Chemical Facilities National Emphasis Program (CPL 03-00-014, Nov. 29, 2011)

Environmental Compliance Guidebook, Beyond US Water Quality Regulations, by Christine Herndon and Shelly Hemming, provides handling and disposal guidelines, as well as regulatory information, for hazardous chemicals and waste.

Compressed Gases

Description of hazard. Water plant operators, associated maintenance staff, and laboratory technicians are often required to work with compressed gases in treatment plants, maintenance facilities, and associated laboratories. Numerous potential physical and health hazards are associated with compressed gases such as propane, oxygen, nitrogen, argon, chlorine, ammonia, and compressed air. Compressed gases can be stored in cylinders (Figure 6-6), portable tanks, or stationary tanks.

Compressed gases are hazardous because they are contained in high pressure containers and can be released accidentally from a broken or leaking valve or from a safety device. When unsecured, uncapped cylinders can be knocked over, breaking the valve and releasing the high-pressure gas, turning the damaged cylinders into uncontrolled rockets or pinwheels, and potentially causing severe injury and damage. Compressed gas in chemical reaction systems can cause vessels to burst if they are improperly released, create leaks in hoses, or produce runaway reactions.

Common controls. All workers handling or using cylinders must have basic training in the use of gas cylinders, emergency shutoffs, proper equipment design, leak-testing procedures, and the use of appropriate respiratory protection in the event of a release of a compressed gas. Other controls include the following:

- In storage, compressed gas cylinders must be restrained using straps, chains, or other suitable stand to prevent them from falling.

Figure 6-6 Properly stacking and restraining pressure vessels reduces potential hazards

- Full cylinders must be segregated from empty cylinders and need to be legibly marked with the chemical or trade name of the gas.
- Storage areas need to be well ventilated to prevent accumulation of explosive concentrations of gas and no ignition sources are permitted in these areas.
- Hand trucks should be used for the transfer of cylinders from storage area to shop or laboratory areas.
- Valve protection covers must be in place during the transport of cylinders. Special procedures are required for changing cylinders of toxic, corrosive, and pyrophoric gases and liquids.
- Never use a cylinder that cannot be identified positively.
- Do not use compressed gas or air to blow away dust or dirt.
- When not in use, close cylinder valves.
- On valves without hand wheels, use the wrench recommended by the gas supplier. The wrench must remain on the valve while the container is in use.
- On valves with hand wheels, never use wrenches or hammer a hand wheel to open or close a valve.

Regulatory information and other resources

OSHA regulations 1910. 251 through 255 for welding, cutting or brazing operations

Canadian Centre for Occupational Safety and Health. www.ccohs.ca/oshanswers/ chemicals/ compressed/compress.html#_1_3

Compressed Gas Association Safe Handling of Compressed Gases Pamphlet P-1 (and other standards). www.cganet.com

Asbestos, Lead, and Mercury

Description of hazard. Lead is most commonly released when heating, cutting or grinding lead piping or gaskets or disturbing old layers of paint. Asbestos is most commonly released when disturbing asbestos high temperature insulation, piping, or flooring. Mercury is most commonly released when old water system manometers, thermometers, instruments and controls are damaged (e.g., float or pressure switches). These substances were present in a surprisingly wide range of other products used in commerce. Residues from release events can remain present for years after the event releasing them due to dust deposition or mercury settling in unobserved locations, such as crevices. Exposure to asbestos, lead, and mercury has serious health implications; these substances primarily affect the nervous system, kidney functions, blood cells, and blood pressure and may have reproductive and developmental effects. Acute (short-term, high level) exposure leads to a wide variety of neurological symptoms including weakness and exhaustion, insomnia, loss of appetite, and kidney disease. Chronic (long-term, high level) exposure can severely affect the gastrointestinal and central nervous systems, leading to hypotension, hearing loss, male infertility, adverse reproductive effects in women, fetal effects, kidney failure, anemia, lethargy, and brain disorders.

Workers can be exposed to these materials through three main routes:

- *Inhalation* occurs when a worker breathes in particulates. Particulates can be generated by a variety of work processes and maintenance tasks. Dusts and/or fumes can be produced from processes such as cutting, welding, and grinding.

- *Ingestion* occurs when a worker consumes products either directly or through cross-contamination from work surfaces or processes to the hands, and then into the body via food, drink, gum, smoking, etc.
- *Absorption* through the skin is a lesser route of exposure but workers should be aware of the potential for hazardous materials to pass through the skin and into the body.

Common controls

Hazard identification. The first step in preventing exposure to hazardous materials is identifying the presence of a potential hazard. Several identification and analytical methods are used:

- Surface test kits
- Usage records
- Lab analysis of samples
- X-ray fluorescence (XRF)
- Purchasing records
- SDS database
- Material/substance inventories
- Awareness training

Risk assessment

Next, evaluate the risk of exposure to the material.

Low-risk activities are those in which workers may be exposed, but the airborne exposure is not likely to exceed the action level for the eight-hour time-weighted average (TWA) exposure limits. Exposures are only expected in rare circumstances as low-risk activities are not likely to disturb products to a degree that would lead to concern.

Moderate-risk activities include actions that directly disturb hazardous materials, through manual or powered methods that may produce airborne exposures and become a health hazard by approaching the eight-hour TWA exposure limits.

High-risk activities are those that may significantly disturb hazardous materials to a degree that airborne particles or fumes are likely to exceed occupational exposure limits.

Exposure controls

Control measures should be considered according to the hierarchy discussed in chapter 4 (elimination, substitution, engineering, administrative, and PPE). Control measures are *not* mutually exclusive, and several methods should be used in concert to control any hazard. Employ controls for higher risk activities in addition to the control measures for lower risk activities (e.g., if the risk activity is moderate, then both low-risk controls *and* moderate-risk controls must be used). Always consult and follow state and local rules governing Asbestos and Lead abatement and licensing before proceeding with anything other than low risk activities. Abatement activities are typically tasked to highly qualified and licensed specialty contractors.

Low-risk activity controls

- Provide washing facilities where practicable, including wash basins, water, soap, and disposable towels.

- Make available appropriate hand cleaning products and disposable paper type towels for field work.
- Prohibit eating, drinking, chewing gum, and smoking in the work area.
- Require hands to be washed before eating, drinking, smoking, or leaving the work area.
- Provide a clean area separate from the work area for coffee and lunch breaks.
- Clean up all dust and waste and place in marked waste containers that are dust-tight.
- Keep the work area clean. Do not use compressed air or dry sweeping to remove dust.

Moderate-risk activity controls

- Erect barriers to prevent access to unprotected workers whenever practicable.
- Secure the area and restrict entry to those with appropriate training and personal protective equipment.
- Wear disposable coveralls if notable amounts of hazardous materials have the ability to contaminate clothing.
- Keep the work area clean using cleaning methods that minimize the generation of dust, such as wet dusting, sweeping, or vacuuming using a vacuum with a HEPA filter on the exhaust.
- Construct partial or full enclosures (if practicable) around work areas where significant scraping, sanding, or demolition will take place.
- Wear a half-face or full-face elastomeric respirator equipped with P100 HEPA cartridges or powered air-purifying respirator (PAPR) equipped with P100 HEPA cartridges. Choose the type of respirator depending on the amount of material removed and the duration of the work.

High-risk activity controls. Do not allow workers to engage in the regular or routine work activities that are associated with significant known exposure potential. Should high-risk work activities be identified, management needs to address the issue during the project planning stage and prior to commencement of the activity.

Regulatory information and other resources

Asbestos, Lead, and Mercury are regulated by OSHA 1910 Subpart Z, Toxic and hazardous substances and state asbestos and lead regulations.

The Agency for Toxic Substances and Disease Registry, a branch of the CDC, has a Toxic Substances Portal which details important information about toxic substances and how they affect our health: www.atsdr.cdc.gov/substances/index.asp

Silica

Description of hazard. Silica, also called silicon dioxide, exists in many different forms and is one of the most common minerals on our planet. Silicosis is an incurable, often fatal, disease caused by exposure to one of the most common forms of silica—crystalline silica dust. Crystalline means that the oxygen and silicon atoms are arranged in a specific pattern. The dust particles are so small you can see them only with a microscope.

Exposure can occur when workers cut, saw, drill, and crush concrete, brick, ceramic tiles, rock, and stone products; and in operations that process or use large quantities of sand, such as foundries and the glass, pottery, and concrete products industries.

Exposures can also occur during construction tasks that require masonry saws, grinders, drills, jackhammers, and hand-held powered chipping tools, vehicle-mounted drilling rigs, milling machines, and heavy equipment for demolition.

Breathing in very small ("respirable") crystalline silica particles causes multiple diseases, including silicosis, an incurable lung disease that leads to disability and death. Respirable crystalline silica also causes lung cancer, chronic obstructive pulmonary disease (COPD), and kidney disease. Exposure to respirable crystalline silica is related to the development of autoimmune disorders and cardiovascular impairment. These occupational diseases are life-altering and debilitating disorders that annually affect thousands of workers across the United States.

Control Elements

- Exposure Control Plan—Prepare a written exposure control plan that describes the tasks that involve exposure to respirable crystalline silica and a description of control and work practices, respiratory protection used for each task, and a description of the housekeeping measures used.
- Workplace Exposure Assessment—Assess the exposure of each employee who is exposed to respirable crystalline silica at or above the action level of 25 $\mu g/m^3$ (micrograms of silica per cubic meter of air), averaged over an eight-hour day. This does not apply to construction-industry employers if they follow the requirements in OAR 437-002-1057, Specified exposure control methods.
- Permissible Exposure limit—Ensure that no employee is exposed to an airborne concentration of respirable crystalline silica that exceeds the permissible exposure limit (PEL) of 50 $\mu g/m^3$ (calculated as an eight-hour time-weighted average).
- Restricted Areas—Restrict employees' access to areas where they could be exposed above the permissible exposure limit. Construction activities require written procedures to restrict access to these areas and a competent person to ensure the procedures are followed.
- Engineering and Work Practice Control—You must use engineering and work practice control methods to keep employees' exposure to respirable crystalline silica at or below the PEL and to reduce exposures as much as possible.
- Construction Activities—Use specified exposure control methods, per state regulations, instead of conducting an exposure assessment including common construction tasks with dust control methods so that employers know what they need to do to limit worker exposures.
- Respiratory Protection—Provide respirators to workers when dust exposures cannot be kept at or below the PEL, when otherwise required by the silica rules, and when required by a task requiring exposure control methods. Respirators must meet the requirements in this rule and the requirements in 1910.134, Respiratory Protection, including the medical evaluation requirement of that standard.
- Housekeeping—Prohibit dry sweeping and brushing when wet sweeping or HEPA-filtered vacuuming are feasible and prohibit use of compressed air to clean surfaces unless the compressed air is used with a ventilation system that captures the dust.
- Medical Surveillance—Offer medical exams, including chest X-rays and lung function tests, every three years for workers who are exposed to silica at or above the action level for 30 or more days per year and for workers that are required to wear a respirator for 30 or more days per year.

- Information and Training—Ensure that each employee has access to labels on containers of crystalline silica and safety data sheets and understands the requirements in 1910.1200, Hazard communication. Employees must also know the tasks and health hazards associated with respirable crystalline silica and the methods used to protect them.
- Recordkeeping—Keep records of all exposure measurements and objective data used to assess employee exposures to respirable crystalline silica. Also, keep records of employees' medical exams if they are under medical surveillance.

Source: Oregon OSHA, https://osha.oregon.gov/OSHAPubs/factsheets/fs67.pdf

Regulatory Information

OSHA 29 CFR 1926.1153 and 1910.1053

Hazardous Waste

Description of hazard. Sudden injuries or illnesses, some of which may be life-threatening, can occur when employees are exposed to safety or health hazards in hazardous waste operations. Activities where hazardous conditions may exist include clean-up operations, corrective actions, treatment, storage, disposal operations, and emergency response. For instance, attempting to clean up a spill of a toxic water treatment chemical, employees may be exposed to harmful toxins in the air and can suffer significant damage to their health; some of it may even be permanent or life-threatening.

Common controls. Proper hazardous waste management is crucial for minimizing risks to employee health and the environment. Employers must properly characterize all waste streams to determine which ones may meet the criteria for being classified as hazardous waste (ignitability, corrosivity, reactivity, toxicity, or listed). For cleanup, site characterization, analysis, and control are required to identify specific hazards, to determine the appropriate health and safety control procedures needed to protect employees or contractors from identified hazards, and to control the hazardous exposures before clean-up begins.

A hazardous waste generator must obtain a USEPA identification number and know its generator status (very small, small or large quantity) in order to implement the specific storage, employee training and other regulatory requirements. Hazardous waste can only be moved to a permitted offsite treatment, storage and disposal facility by a licensed hazardous waste transporter using a hazardous waste manifest to track it to its final destination and fate.

Employers who have staff working in hazardous waste cleanup operations and performing active emergency response must meet higher standards requiring a written safety and health program to identify, evaluate, and control the safety and health hazards and provide training on recognizing hazards, properly using PPE, understanding work practices that will reduce risks, using of engineering controls and equipment, and recognizing the signs which may indicate overexposure to a hazard.

Regulatory information and other resources

OSHA Hazardous Waste Operations and Emergency Response standard is 29 CFR 1910.120

USEPA standards for hazardous waste management, including identification, storage and handling, can be found in 40 CFR 263.

The handling, transportation, communications, emergency response, training requirements, and security plans for hazardous materials can be found in 49 CFR 171–179.

Environmental Compliance Guidebook—Beyond US Water Quality Regulations, by Christine Herndon and Shelly Hemming, provides guidance in developing hazardous waste management programs for complying with federal environmental requirements.

Pesticides (Licensed Applicators)

Description of hazard. Water plant operations and maintenance tasks sometime involve the application of pesticides. These work activities include the handling of pesticides. Workers must be properly trained and licensed to handle these products. Storage of pesticides and empty containers must not be in proximity to water wells or other water treatment facilities to avoid accidental contamination or vandalism.

Common controls. Regulatory programs have developed specific rules for pesticide safety, including information on about the following:

- Pesticide selection
- Handling, mixing, and loading pesticides
- Application of pesticides
- Re-entry into treated fields and required posting
- Protective clothing and equipment and its care and maintenance
- Pesticide poisoning effects, symptoms, and remedies
- Medical supervision for pesticide applicators
- Toxicity of pesticides
- Minimizing exposure
- Risk assessment
- Personal safety and hygiene
- Types of spray equipment
- Equipment parts, with emphasis on nozzles
- Sprayer maintenance and cleaning
- Sprayer storage
- Pesticide transfer — closed systems

The licensed applicator should be made familiar with the various types of hand, ground, and aerial equipment available and with the various parts and functions of spray equipment. The applicator must be skilled in sprayer equipment calibration, maintenance, cleaning, and storage. The applicator should know that proper calibration of spray equipment is one of the most important factors in pesticide application equipment and to understand the importance of performing frequent calibration.

To protect all workers and the general public from pesticide exposure, licensed applicators must oversee all aspects of transporting, storage, and disposal of pesticides. The following activities must be carefully planned and implemented as needed:

- Emergency response to spills and fires
- Bulk storage and handling, disposal and reuse of dilute (left over) spray mixture
- Pesticide record keeping

Workers must be trained in the proper decontamination techniques and the approved disposal alternatives for empty pesticide containers and unused or unwanted portions of pesticides.

Regulatory and other resources:

Standards for certification of applicators of restricted use pesticides were first printed in the *Federal Register* on October 9, 1974. The general standards from which these were derived are found in the Code of Federal Regulations, 40 CFR Parts 171.1–171.6 (categories for minimum standards) and 40 CFR Parts 171.7–171.11 (regulations for state plans for implementing certification).

Federal Worker Protection Standard, 40 CFR Part 170, covers pesticides Local, state, provincial, and federal rules related to right-to-know laws, endangered species, pest resistance (local problems), and chemical/irrigation issues.

NIOSH training materials are available at www.cdc.gov/niosh/topics/outdoor/.

Iowa State University Extension and Outreach website www.extension.iastate.edu/workerprotection/.

Environmental Compliance Guidebook: Beyond US Water Quality Regulations provides guidance in developing pesticide management programs for complying with federal environmental requirements.

PERSONAL (OR HEALTH) HAZARDS

Drug and Alcohol Abuse in the Workplace

Description of hazard. The harmful use of drugs, smoking/vaping, and alcohol in the workplace is not only a vital safety concern, it adds to the cost of injuries, absenteeism, lost production, worker compensation, and rehabilitation. It is estimated that the direct cost of sickness and death from drug and alcohol abuse to industry is $3.7 billion per year.

Harmful use of drugs and alcohol can create a range of problems in the workplace. Employees with drug and alcohol problems can cause injury to themselves and others, jeopardize their job, and damage their physical and mental health.

Two kinds of drinking behavior and drug abuse significantly contribute to work-performance problems. One behavior is abusing alcohol or illegal/nonprescription drugs right before or during working hours (during lunch or company functions). The other is drinking alcohol or using illegal/nonprescription drugs the night before, causing hang-overs during work the following day.

In the workplace, the impact of alcohol and drug abuse affects four major areas: accidents, productivity, medical expenses, and employee morale.

Accidents. Operating machinery under the influence of alcohol or drugs increases the likelihood of accidents and injuries. The American Council for Drug Education reports that substance abusers are about three and a half times more likely to be involved in accidents on the job and five times more likely to hurt themselves at work. They are also five times more likely to file for worker's compensation. Substance abusers are also responsible for 40 percent of all industrial fatalities. OSHA reports that 10 to 20 percent of US workers who die at work test positive for drugs and alcohol.

Productivity. Employee drug and alcohol use results in lost efficiency and quality, because substance-abusing employees are 33 percent less productive when they are at work than those who are sober, according to the American Council for Drug Education. These employees also increase human resource costs related to additional recruitment and training because substance abusers are 10 times more likely to miss work. Replacing an employee can cost anywhere from 25 to 200 percent of that employee's annual compensation.

Medical expenses. Employee substance abuse results in higher health care expenditures for illnesses and injuries, increasing premium costs. US Department of Health and Human Services studies show that drug abusers use twice as many medical benefits as their coworkers and are directly responsible for health care costs that are three times higher than the national average.

Employee morale. An organization that tolerates substance abusers runs the risk of losing conscientious and hardworking staff. Over time, the best workers will tire of working harder while substance abusers produce less and enjoy more time off. An estimated 3.1 percent of employed adults admit they have used illicit drugs before reporting to work and 7.1 percent say they drank on the job one or more times in the past year, according to sources cited by the US Department of Labor.

Additional problems from substance abusers can include: tardiness and sleeping on the job, theft, poor decision making, increased likelihood of having trouble with coworkers and supervisors or tasks, higher turnover, and disciplinary procedures.

Common controls. Work can be an important and effective place to address alcoholism, smoking/vaping and other drug issues by establishing or promoting assistance policies, training, and programs. Back in the late 1940s and early 1950s, the National Council on Alcoholism and Drug Dependence (NCADD) found that the workplace was ideally suited to address alcoholism through a focus on employee job performance and access to treatment. NCADD founded the employee assistance program (EAP) concept—then known as Operational Alcoholism Programs—as a joint labor-management program and wrote the first EAP manual and standards.

Without question, establishment of an in-house EAP is the most effective way to address resources and services. By encouraging and supporting treatment, employers can dramatically assist in reducing the negative impact of alcoholism and drug abuse in the workplace, while reducing related costs. Research has demonstrated that treatment programs pay for themselves in reduced health care costs that begin as soon as people begin treatment. Treatment programs also improve an individual's ability to function, leading to increased productivity at work.

In short, the steps to reducing the effects of drug and alcohol abuse in the workplace are as follows:

1. Establish and enforce policies again drug and alcohol use.
2. Train supervisors and staff on recognizing impaired behavior and signs of alcohol/drug use.
3. Conduct employee testing when employee exhibits signs of impairment of drug/alcohol use.
4. Implement random testing procedures.
5. Establish an EAP.

Regulatory information and other resources

National Council on Alcoholism and Drug Dependence, Inc. has information about drugs and alcohol in the workplace at ncadd.org.

Centers for Disease Control and Prevention. www.cdc.gov

Ergonomic Hazards

Description of hazard. Ergonomics is the science of fitting workplace conditions and job demands to the capabilities of the worker. Effective and successful "fits" help ensure high productivity, reduce illness and injury risks, and increase satisfaction among the workforce. Although the scope of ergonomics is much broader, the term here refers to assessing those work-related factors that may pose a risk of musculoskeletal disorders (MSDs) and recommendations to alleviate them.

Vibration and cold may add risk to these work conditions. Jobs or working conditions presenting multiple risk factors will have a higher probability of causing a MSD. The level of risk depends on the intensity, frequency, and duration of the exposure to these conditions. Environmental work conditions that affect risk include intensity, frequency, and duration of activities.

Common controls. Several approaches may be used individually or in combination to determine whether conditions in the workplace might be contributing to employees developing MSDs.

Review and analyze injury and illness records, including OSHA 300 Logs and supporting 301 forms and workers' compensation claims, to determine whether there is a pattern of ergonomic-related injuries in certain jobs or work tasks.

Analyze the jobs or work tasks to identify potential ergonomic problems before employee injuries occur. Determine if jobs present ergonomic risks that may contribute to MSDs. Much of this work can be done in conjunction with the job safety analyses discussed in chapter 2.

Be aware of common contributing conditions within the industry or job classifications. If other water organizations have ergonomic-related problems, then these problems may be common throughout the industry and should be looked at in each facility or utility.

A wide variety of possible solutions can be implemented to reduce or eliminate the ergonomic risk associated with jobs or work tasks in the water treatment industry.

Regulatory information and other resources

OSHA Ergonomic eTools are stand-alone, interactive, web-based training tools are available at https://www.osha.gov/SLTC/etools/computerworkstations/index.html

OSHA publications, *Job Safety and Health Quarterly* articles, and information from other agencies and sources.

First Aid and Access to Medical Assistance

Description of hazard. Sudden injuries or illnesses do occur in the workplace. Without immediate intervention, some injuries or illnesses may be life-threatening. Sudden cardiac arrest (SCA) from asphyxiation, electrocution, or exertion may occur. Many water system employees are located far from medical facilities, and the response time for emergency medical services (EMS) could be long.

Common controls. Designated staff should be trained as first-aid providers at all workplaces of any size if there is no medical facility nearby. Training in cardiopulmonary resuscitation (CPR) is particularly important because of the possibility of SCA. CPR may keep a victim alive until EMS arrive to provide the next level of medical care. An onsite automatic external defibrillator (AED) further increases the chances for survival.

Regular first-aid training is just one part of a workplace first-aid program, which should also include first-aid kits in accessible locations, including vehicles, a designated person who stocks and maintains the first-aid kits, knowledge of the response times for the local fire and rescue service or emergency medical professionals, written policies and procedures, and management commitment and worker involvement.

The first-aid program should reflect the known and anticipated risks of the workplace, with specific response training and supplies for injuries and illnesses that are known to occur in similar work environments, many of which are discussed in this manual.

A well-stocked first-aid kit includes

- abdominal pad;
- individually wrapped adhesive dressings (sterile gauze pads, rolls of gauze bandage, field dressings, triangular bandages);
- antiseptic wipes;
- aspirin and non-aspirin for pain;

- burn cream;
- cold pack;
- dressing;
- eye wash;
- first-aid cream;
- first-aid manual;
- latex gloves;
- safety pins;
- scissors;
- splints and splint padding;
- sterile bandage compresses; and
- tweezers.

Regulatory and other resources

The OSHA Recording and Reporting Occupational Injuries and Illnesses regulation, 29 CFR 1904 provides specific definitions of *first aid and medical treatment.*

A specific example of the minimal contents of a workplace first aid kit is described in American National Standards Institute ANSI Z308.1-2015, *Minimum Requirements for Workplace First Aid Kits.*

OSHA *Best Practices Guide: Fundamentals of a Workplace First-Aid Program,* https://www.osha.gov/Publications/OSHA3317first-aid.pdf

OSHA standard on medical services and first aid, 29 CFR 1910.151

Occupational Exposure to Bloodborne Pathogens standard 29 CFR 1910.1030OSHA standards requiring CPR training: Permit-required Confined Spaces,1910.146; Qualifications of Dive Team, 1910.410; and Construction Subpart V, Power Transmission and Distribution, 1926.950

Lone Worker

Description of hazard. Working alone can be more hazardous than performing the same job when other people are present. A lone worker is defined as an employee who performs an activity that is intended to be carried out in isolation from other workers, without close or direct supervision. Such staff may be exposed to risk because there is no one to assist them should they be injured. If a lone worker is injured or an emergency occurs the worker will not have access to help if no one is within shouting distance. If the worker is rendered unconscious, it may be hours before help arrives, and by then it may be too late.

Common controls. The following are safety recommendations for a variety of situations for lone workers:

- Conduct a risk assessment of the job activity to determine if it can be done safely by a lone worker under most circumstances.
- Communicate with employees about the specific risks they face and how to minimize them. If work cannot be rescheduled and must be performed alone, treat the tasks the same as any other, with a job safety analysis that clearly spells out the risks.
- Require those who work alone to carry a cell phone, two-way radio, or personal alarm.

- Implement a call-in system where cell phones or two-way radios are used to communicate with co-workers at regular intervals. This may be adequate in low-risk working alone situations and can be augmented by before/after call in for higher risk tasks.
- Require lone workers to file a travel plan that describes their anticipated destinations.
- Ensure that foot travel paths, especially those traveled after sundown, are free of potential trip and fall hazards and that they are properly illuminated.

OSHA's underground construction rule for tunnels, shafts, chambers, and passageways states: "Any employee working alone underground in a hazardous location, who is both out of the range of natural unassisted voice communication and not under observation by other person, shall be provided with an effective means of obtaining assistance in an emergency." Although no other federal OSHA rules specifically apply to working alone, the broad requirements of the safety and health regulations still apply.

Regulatory and other resources

OSHA regulation 29 CFR 1926.8000 for Underground Construction Rule for Tunnels, Shafts, Chambers and Passageways

Blackline GPS is a safety tracking system provider http://www.blacklinegps.com.

Respiratory Hazards

Description of hazard. Water plant operation and maintenance staff may be exposed to various respiratory hazards, including dusts, chemical vapors, and gases. When engineering controls, such as general and local exhaust ventilation systems, are not adequate to control the hazard, workers must be required to use respiratory protection. Emergency procedures may also require the use of emergency escape or self-contained respiratory protective equipment.

Common controls. If an assessment of the work environment and assigned job tasks indicate that workers may be present in an environment above established OSHA or state exposure limits, a written respiratory protection program must be developed and followed. A worksite-specific respiratory program includes

- procedures for selecting respirators;
- medical evaluation to ensure fitness to wear a respirator;
- fit-testing protocol;
- training in respiratory hazards, respirator use, cleaning, storage, and inspection; and
- procedures and schedules for cleaning, disinfecting, storing, inspecting, repairing, discarding, and otherwise maintaining respirators as well as procedures. For atmosphere-supplying respirators (SCBA and airline supplies), procedures to ensure adequate air quality, quantity, and flow of breathing air.

The written respiratory protection program must be communicated to all affected employees. A sample request for medical clearance for respirator use and medical and a sample respirator use and maintenance record form should be included in the written program.

Training in respiratory protection is critical to worker protection and should integrate clear toxic gas concentration thresholds for donning a respirator as well as the Maximum Use Threshold (MUC) based on either manufacturer provided service limitations or the Immediately Dangerous to Life and Health (IDLH). Retraining is required annually or whenever changes in the type of respiratory protection occur. Only NIOSH-certified respirators can be used in the workplace.

Regulatory information and other resources

OSHA regulation on respiratory protection 1910.134
NIOSH training materials are available at www.cdc.gov/niosh/topics/

Sanitation and Disease Prevention

Description of hazard. Many harmful infectious agents are potentially present in raw water supplies, along waterways and in waste materials generated by water treatment operations. These infectious agents include a variety of viruses, bacteria, and various parasites. Working outdoors also introduces exposure to vector-borne diseases, including Lyme Disease, West Nile Virus and other mosquito-borne diseases, along with various zoonotic infection, such as leptospirosis. Open water reservoirs also attract waterfowl and wildlife that may carry zoonotic pathogens described above under waterborne hazards.

Common controls. Maintaining clean and sanitary conditions is a primary concern in the water industry and ensures a workplace that protects the public health. General housekeeping requirements, water supply, toilets, washing facilities and lavatories, locker rooms, and eating and food storage areas are all covered by workplace health and safety regulations.

Floors need to be maintained in a clean and dry condition to avoid worker injuries from slips, trips, or falls. Waste receptacles should be maintained in a sanitary condition and all garbage and refuse should be removed daily to prevent the entrance or harborage of rodents, insects, and other vermin. Nonpotable water inside a facility must be posted or marked to indicate the water is unsafe for drinking, personal cleaning, cooking, or washing clothes. Food and beverages should not be stored in toilet rooms or in any area that is exposed to toxic chemicals, such as laboratory refrigerators or coolers.

If workers are potentially exposed to vector-borne diseases spread by infected ticks and mosquitoes, they should use insect repellent and wear long-sleeved shirts and long pants that can be tucked into boots or socks. After potential exposure, skin and clothing need to be checked for ticks, which should be removed using tweezers. Anyone who develops symptoms of a tick-borne disease should seek medical attention promptly. In warmer months, working outdoors during dawn and dusk should be avoided if mosquitoes are present.

Regulatory information and other resources

OSHA regulations on sanitation, 1910.141
NIOSH training materials are available at www.cdc.gov/niosh/topics/

Violence in the Workplace

Description of hazard. Workplace violence is violence or the threat of violence against workers. It can occur at or outside the workplace and range from threats and verbal abuse to physical assaults and homicide, one of the leading causes of job-related deaths. However, it manifests itself, workplace violence is a growing concern for employers and employees nationwide.

Common controls

Hiring process. Efforts to avoid workplace violence start with the hiring process. Organizations should require all hiring managers to follow guidelines for screening job applicants in an effort to reveal problematic behavior that could create an unsafe workplace. Employees should also participate in preventing violence by alerting their employers about potentially harmful behavior demonstrated by co-workers.

Zero tolerance policy. Reducing and preventing workplace violence starts at the top. Management must demonstrate its commitment by sponsoring and funding a violence-free workplace program, including a zero-tolerance policy. The policy should prohibit any violence by anyone on the company property or involved in any business-related activities. It should include reporting, investigation, and discipline procedures, along with a confidential communication process so employees can report threats or acts of violence without fear of retaliation. All employees and managers should receive training on the policy.

Conduct a workplace violence risk assessment to identify potentially violent situations. Take a look at environmental conditions, such as physical conditions at the work site; organizational issues, such as recent restructuring or workforce reductions; drug and alcohol use in the workplace; and bullying or intimidation. Evaluate each potential risk and develop a plan to address those identified as high risk and those most likely to occur or cause serious injuries.

Workplace safety. Make it clear to employees that everyone is responsible for workplace safety. Encourage employees to report to supervisors disturbing behavior that could lead to violence among co-workers. This includes incidents of a worker harassing, stalking, or threatening another worker. Companies that require employees to wear photo-identification badges or carry key cards to enter officers should establish security rules. For example, employees should be aware of disciplinary actions they might face if they lend their key cards to people who are not authorized to enter company offices. Such actions could allow a dangerous person to enter the workplace.

Employee conflict. Supervisors should be required to set an example for appropriate workplace behavior. They must make it clear to subordinates that threatening behavior in the workplace will not be tolerated. Supervisors need to promptly address conflicts between co-workers that could turn violent, especially if the same employees are the source of the conflicts. A company's human resources department should be ready to advise supervisors on appropriate measures to take to address conflicts and head off potentially violent situations among employees.

The Crisis Prevention Institute recommends these tips for preventing workplace violence:

Assess the work environment. Critically examine all areas of the work environment, including parking lots, entryways, reception areas, work areas, and offices for safety. Is the lighting adequate? Are there convenient escape routes? What is the method to summon assistance?

Pay attention to the warning signs. Many people who become violent communicate their intentions in advance. Threats from customers, coworkers, or third parties should be reported immediately.

Promote respect. The best way to prevent violence in the workplace is to foster a day-to-day attitude of respect and consideration in the work environment.

Eliminate potential weapons. Take a mental inventory of objects available in the immediate work area that could be potential weapons. Remove or secure objects that could be thrown.

Know violence response procedures. Violence response procedures are simple plans designed to minimize injury during a violent incident. These procedures should include a plan to summon assistance and move people to a safe area.

Trust your instincts. Do not ignore your internal warning system. If you sense impending danger, react accordingly.

Employ the buddy system. If you are in a situation in which hostility could occur, use the buddy system, so you are not alone in places where violence could occur.

Other resources

CDC has Occupational Violence Resources available at www.cdc.gov/niosh/topics/violence.

Crisis Prevention Institute provides training and literature in workplace violence awareness and prevention. www.crisisprevention.com

ADDITIONAL INFORMATION

Additional information specific to safety during emergencies can be found in AWWA Manual of Water Supply Practices, M19, *Emergency Planning for Water Utilities*. To ensure a full understanding of the hazards, further research is recommended. Also, additional topic-specific safety tips are provided for weekly safety meetings in AWWA's annual publication *Let's Talk Safety: 52 Talks on Common Utility Safety Practices.*

The OSHA website, www.OSHA.gov. OSHA has various publications, standards, technical assistance, and compliance tools, and offers extensive assistance through its many safety and health programs: workplace consultation, voluntary protection programs, grants, strategic partnerships, state plans, training, and education. Guidance such as *OSHA's Safety and Health Management Program Guidelines* identify elements that are critical to the development of a successful safety and health management system.

In states with their own programs (see www.osha.gov/dcsp/osp/index.html) utility managers should review the state requirements as sometimes they differ from OSHA requirements. For Canadian utilities, the links listed in the Appendix A: Canadian Content should be used. Also, consider contacting AWWA Section Safety Committees to inquire how others are addressing various hazards.

Appendix A contains guidance documents and forms to show readers what other water agencies are using to protect employees and contractors.

Appendix **A**

Guidance Documents and Forms

The following guidance documents and forms may be valuable in ensuring a safe workplace for employees and contractors in the water industry.

AWWA M3 SAFETY MANAGEMENT FOR WATER UTILITIES

Accident Investigation Program

These procedures are to be followed by the victim [you] of an accident or near miss.

I. Program

In order to prevent recurrences, the Utility investigates all workplace accidents. All accidents, regardless of severity, will be analyzed to determine the best method of prevention. Listed below are procedures that are to be followed after an accident. These procedures will ensure that you are taken care of after being involved in an accident, all pertinent information is gathered, and that all preventive measures are followed through.

II. Procedures

A. **Near Miss**

A near miss is an incident where no injury occurred, but under slightly different circumstances, could have involved injury/injuries.

1. You shall report to your supervisor by the end of the workday.
2. If possible, remove the hazard or take action to prevent reccurrence immediately.
3. You and your supervisor will complete a near-miss form (forms are located ____).
4. Turn a copy of the near-miss form into the Safety Coordinator by the next workday for review (keep a copy to review with your crew).
5. Review the near miss and actions taken at your next crew meeting. Turn in a copy of the meeting minutes to the Safety Coordinator.
6. The Safety Coordinator will review all information with the Safety Committee.
7. Your Safety Committee representative will report any recommendations from the Safety Committee at your next crew meeting.

B. **First-Aid Injury**

A first-aid injury is a minor injury that can be treated on site and does not cause days away from work or a job transfer.

1. You shall report to your supervisor by the end of the workday.
2. If possible, remove the hazard or take action to prevent reccurrence immediately.
3. You and your supervisor will complete an injury report form (forms can be found __________________________).

Page 1 of 3

Courtesy of Tualatin Valley Water District

Figure A-6 Accident investigation program

AWWA M3 SAFETY MANAGEMENT FOR WATER UTILITIES

Accident Investigation Program

4. Turn a copy of the injury report form into the Safety Coordinator by the next workday for review (keep a copy to review with your crew).
5. The Safety Coordinator will meet with you to review the incident and evaluate the status of your injury.
6. The Safety Coordinator will look for similar injuries incurred while performing the same task, that occurred within the past 5 years. If a similar injury is found, a root-cause analysis must be done.
7. Review accident at your next crew meeting. Turn in a copy of the meeting minutes to the Safety Coordinator.
8. The Safety Coordinator will review information with the Safety Committee.
9. Your Safety Committee representative will report any recommendations from the Safety Committee at your next crew meeting and turn in a copy of the meeting minutes to the Safety Coordinator.

C. **Major Injury**

An injury that requires professional medical attention, days away from work, or job transfer is considered a major injury.

1. If necessary, call 911.
2. Report to your supervisor as soon as possible.
3. If an ambulance is not necessary, your supervisor will transport you to the nearest medical facility (a post-accident drug and alcohol test will be requested).
4. Your supervisor will notify the Safety Coordinator.
5. The Safety Coordinator will initiate investigation of the accident, start the 801 form, and notify the Utility's Worker's Compensation carrier.
6. As soon as possible, you will be asked to provide information for the injury report and the 801 form.
7. If the injury results in admission to a medical-care facility for medical treatment, the Safety Coordinator will report to Occupational Safety and Health Administration (OSHA) within 24 hours.
8. A root-cause analysis will be done within two (2) days of the accident. The Safety Coordinator and the Emergency Planning, Risk & Security Coordinator will conduct this analysis.
9. Based on the root cause(s) identified, recommendations will be made to prevent a reccurrence.
10. The Safety Coordinator will bring the results of the analysis, along with recommended actions, to the Safety Committee.
11. The Safety Committee representatives will share the information with their departments.

Page 2 of 3

Courtesy of Tualatin Valley Water District

Figure A-6 Accident investigation program (*continued*)

AWWA M3 SAFETY MANAGEMENT FOR WATER UTILITIES

Accident Investigation Program

D. **Fatality/Catastrophe**

If an accident results in a fatality or a catastrophe [two (2) or more fatalities or three (3) or more admitted to a medical-care facility] the following will be done.

1. The Safety Coordinator and/or the Human Resources Manager will contact family members.
2. The Safety Coordinator will contact OSHA within 8 hours.
3. An investigation and root-cause analysis will be conducted (see Major Injury steps 8–11).

E. **Vehicle Accident**

1. If you or anyone else involved in the accident are injured, assess injury/injuries and call 911, if necessary.
2. Police response is required if a fatality occurs, an injury requires transport by ambulance, a person involved appears to be under the influence of intoxicants, or the accident involved a hazardous materials spill.
3. Notify your supervisor as soon as possible after the accident.
4. Supervisor will notify the Safety Coordinator and/or the Emergency Planning, Risk & Security Coordinator and coordinate a response.
5. You and/or the response team will gather all necessary information. Follow the procedure defined in the Utility Driving Policy.
6. Turn in the Accident Report and Department of Motor Vehicles (DMV) report by the end of the shift.
7. DMV report will be submitted to DMV within 72 hours.
8. Drug and alcohol testing will occur according to Driving and/or Substance Abuse policies.
9. The Emergency Planning, Risk & Security Coordinator will turn information into the Utility's insurance carrier.
10. Accident will be investigated to determine if it was preventable.
11. If the accident is deemed preventable, your DMV record will be evaluated per the Driving Policy.
12. Findings and recommendations will be shared with you and the Safety Committee.
13. Safety Committee representatives will share information with their groups.
14. If you are injured, the severity of the injury will be determined and the procedures outlined above will be followed.

Page 3 of 3

Courtesy of Tualatin Valley Water District

Figure A-6 Accident investigation program (*continued*)

AWWA M3 SAFETY MANAGEMENT FOR UTILITIES

☐First Aid
FILE 801, IF BOXES BELOW ARE CHECKED
☐Medical Care
☐Time Loss
☐Fatal

SYSTEM CHALLENGES

Management
Policy Enforcement
Hazard Recognition
Accountability
Supervisor Training
Corrective Action
Production Priority
Proper Resources
Job Safety Training
Hiring Practices
Maintenance
Adequate Staffing

Employee
Following Procedure
Training
Previous Injury
Mental Ability
Physical Capacity
Equipment Use
Short Cuts
PPE Worn
Safety Attitude

Equipment
Proper Tool Selection
Tool Availability
Maintenance
Visual Warnings
Guarding

Environment
Plant Layout
Chemical
Temperature
Noise
Radiation
Weather
Terrain
Vibration
Ergonomics
Lighting
Ventilation
Housekeeping
Biological

Additional Causal Factors:
☐Faulty Equipment
☐Non-Employee
☐Prior Injury
☐Late Reporting
☐Off-the-Job Injury
(Describe checked boxes on another sheet)

Accident/Incident Analysis

Immediate supervisor should complete this form promptly with affected worker.

Employee:____________________

Occupation/Department:____________________

Where Incident Occurred:____________________

Date:________________ Time:__________ am☐ pm☐

If injury, describe (nature/body part):____________________

Treatment: ☐ None ☐ First Aid Only ☐ Doctor ☐ Hospital

Treating Physician:____________________

Treating Facility:________________ Phone:________________

Witnesses:____________________

Describe Accident/Incident Fully:

If this injury required first aid, please fill out fields below.

Identify factors that contributed to or caused accident (refer to list on left):

Management:	**Employee:**
Equipment:	**Environment:**

Countermeasures/best practices to prevent recurrence:	
By Whom:	**By When:**

(Attach additional sheet if needed)

Safety Committee Review Date:____________________

If accident/incident was caused by a person not employed by us, who?
Name: ________________ Phone:________________

Date: ________ Supervisor's Signature________________

Courtesy of Tualatin Valley Water District

Figure A-7 Accident/incident analysis

AWWA M3 SAFETY PRACTICES FOR UTILITIES

[Insert Water Utility Name]

Title: Conservation Coordinator **Reporting to: Manager, Office of Community & Intergovernmental Relations (OCIR)**	**Job Grade: 7** **Department: OCIR**

[Insert Water UtilityName] believes that each employee makes a significant contribution to our success. That contribution should not be limited by the assigned responsibilities. Therefore, this job description is designed to outline primary duties, qualifications and job scope, but not limit the incumbent nor the organization to just the work identified. It is our expectation that each employee will offer his/her services wherever and whenever necessary to ensure the success of our endeavors.

Overall Purpose of the Job: The Conservation Coordinator supports the implementation of the Utility's water conservation goals by developing, coordinating, implementing and monitoring Utility conservation programs. This position is responsible for ensuring that the Utility's programs are consistent with Utility, Regional and State requirements and regulations. Specifically, this position will include: planning and implementing Utility conservation programs and events; evaluating and making recommendations to improve programs and events; compiling and analyzing data regarding the Utility's conservation programs. This position is part of a team of employees working to develop conservation opportunities, information and tools.

Essential Job Functions:

- Develop, measure and implement residential conservation program; recommend changes to ensure efficiency and effectiveness of the program.
- Develop a schedule for marketing of Utility conservation programs and activities, including public presentations.
- Provide conservation assistance to residential customers.
- Actively participate and assist in the coordination and organization of Utility conservation-related events.
- Coordinate activities with other related internal and external programs.
- Provide budget recommendations to the Manager of the Office of Community and Intergovernmental Relations (OCIR); assist in tracking, reporting and monitoring the conservation portion of the budget.
- Research and develop additional conservation programs for consideration.
- Represent the Utility at local, regional and state conservation activities and committees.
- Coordinate with regional partners to enhance the effectiveness and ensure consistent messages of both the Utility's and Region's conservation programs, goals and work plans.
- Monitor federal, state and local requirements and regulations. Collect, analyze and report on compliance of the conservation program. Make recommendations based on findings.
- Participate in development of water conservation related material for distribution to schools, businesses and residential water users.
- Serve as a resource to the community and government organizations regarding water conservation programs.
- Provide information, direction and assistance to customer service representatives and other Utility staff to facilitate delivery of water conservation information to customers and the public.
- Participate in collection and analysis of water consumption data for trends and forecasting of water needs for the Utility.
- Coordinate the updating of the Utility's Demonstration Garden, including all informational materials, kiosk, signage and promotion of the Garden and events.
- Implement special projects and other duties as assigned by the OCIR Manager.
- Driving is a job function, your driving record must comply with the Utility's driving policy.

Courtesy of Tualatin Valley Water District

1 of 3

Figure A-8 Sample job description (light physical requirements)

AWWA M3 SAFETY PRACTICES FOR UTILITIES

Knowledge Required:

- Knowledge of the water industry and conservation programs, regulations, policies and procedures at the federal, state and local levels.
- Knowledge of budgeting principles and practices.
- Knowledge in performing research, analysis and report preparation including program evaluation.
- Excellent knowledge of basic computer operations and Microsoft applications such as Word, Excel, Outlook, etc.

Special Skills and Qualities:

- Excellent communication skills, both oral and written, including strong public presentation skills, experience developing reports and community education materials.
- Excellent analytical ability, including the ability to evaluate alternatives and make sound, creative recommendations and present recommendations to individuals from all levels of the Utility, as well as individuals outside the Utility.
- Excellent customer service skills.
- Ability to make decisions independently in accordance with established policies and use.
- Good judgment in carrying out tasks and responsibilities with little supervision or guidance.
- Ability to learn the Utility's customized utility billing computer program and any supporting programs.
- Good organizational skills.

Minimum Requirements:

- Bachelor's Degree with two years of experience in areas that relate to conservation, water use, public relations, or related fields **or** an Associate's Degree with six years of experience in the drinking water industry or water-related profession, **or** equivalent work experience.
- A valid State Driver's License; have and maintain a safe driving record.

Working Conditions:

- Lift, carry and set-up awkward display equipment.
- Work nine hours per day and other hours as required.
- Have regular and predictable attendance.
- Attendance at meetings and events at various locations in and out of the Metro Area, often outside of normal working hours.
- Daily focus on personal computer screen.
- Verbal communication on a daily basis both in person and on the telephone.
- Prolonged periods of sitting or standing.
- Respond and work, without advanced notice, in times of adverse weather, natural disaster, emergency, or other unusual events as determined by the Utility.

Courtesy of Tualatin Valley Water District

2 of 3

Figure A-8 Sample job description (light physical requirements) (*continued*)

[Insert Water Utility Name]

Physical Capacities Requirements

Position: Conservation Coordinator

In terms of a 9 hour workday:

"Occasionally" equals 1% to 33% "Frequently" equals 34% to 64% "Continually" equals 65% to 100%

In a 9 hour workday, the worker will be:

(Total hours at one time)	(Total hours during entire workday)
Standing 0 ½ 1 2 3 4 5 6 7 8 9 Occasionally	Standing 0 ½ 1 2 3 4 5 6 7 8 9 Occasionally
Sitting 0 ½ 1 2 3 4 5 6 7 8 9	Sitting 0 ½ 1 2 3 4 5 6 7 8 9

Job requires repetitive use of hands for:

	Right	*Left*
Simple Grasping	Yes	Yes
Pushing & Pulling	Occasionally	Occasionally
Fine Manipulating	Occasionally	Occasionally

Job requires repetitive movement of feet as in operating foot controls:

Right	*Left*	**Both**
N//A	N/A	N/A

Job requires worker to lift:	Never	Occasionally	Frequently	Continually
Up to 5 lbs.			X	
6 to 10 lbs.		X		
11 to 20 lbs.		X		
21 to 50 lbs.		*Rare, yet 25 lbs might occur		
51 to 100 lbs.	X			

Job requires worker to carry:	Never	Occasionally	Frequently	Continually
Up to 5 lbs.			X	
6 to 10 lbs.		X		
11 to 20 lbs.		X		
21 to 50 lbs.		*See note above		
51 to 100 lbs.	X			

Job requires worker to:	Never	Occasionally	Frequently	Continually
Bend		X		
Squat	X			
Crawl	X			
Reach above shoulder level		X		
Walk or work on uneven surfaces		X		

Job requires the following activities:

	Never	Occasionally	Frequently	Continually
Being around machinery	X			
Operating heavy equipment	X			
Exposure to dust, fumes, and gases	X			

Comments: Will need to be able to set up and take down displays that may be awkward. Will need to be able to work independently with out assistance.

Manager's Signature:____________________________________Date:____________________

Courtesy of Tualatin Valley Water District

Figure A-9 Physical capacities requirements

Purpose

[Insert Water Utility Name] has developed a program to assist workers who temporarily have restrictions due to an injury or illness. This program is called the "Early Return to Work" (ERTW) program.

Policy

Although it is not a requirement that the Utility provide such a benefit, studies show that an Early Return to Work Program is therapeutic and helps speed the recovery process. In addition, injured employees stay in touch with the speedy transition back to their normal job.

I. Reporting

A. To qualify for the Utility's ERTW program, you must report all injuries and illnesses that affect your ability to return to your regular work responsibilities.

B. All injuries/illnesses reported by the employee as work-related and/or filed with the Utility's workers' compensation carrier will be investigated.

II. Work-Related Injury/Illness

A. If you sustain a work-related injury/illness, report it to your supervisor immediately. Your supervisor will first determine the appropriate action to take based on your condition, then report to the Safety Coordinator and ensure that an Injury Report is filed. If the injury or illness requires a visit to a physician, the Safety Coordinator will provide you with the appropriate forms to be given to the treating physician. If the injury or illness requires immediate emergency medical care, the Safety Coordinator will ensure all forms are forwarded to your treating physician.

B. If a visit to a physician is necessary, you must file an 801 Form (Report of Job Injury or Illness) within 24 hours of reporting the injury or illness. If the injury or illness requires immediate emergency medical care, the 801 Form must be filed within 5 days of your visit to the doctor.

C. If you are unable to return to work, you must notify your supervisor of your status the next work day. You must also contact Human Resources to determine if you are eligible for Family Medical Leave.

D. The Utility has contracted with *(Insert Provider Name)* to provide managed care services for Utility workers that are injured on the job. However, you have the right to be treated by your own physician, at your expense, provided your doctor agrees to follow and cooperate with the Utility's Early Return to Work Program. If the injury or illness is determined to be work-related and Workers' Compensation benefits are provided, the Utility's insurance carrier may select a managed care organization. You will be required to seek treatment with a panel-approved medical provider who is familiar and willing to treat occupational injuries. The Utility's insurance carrier will provide an extensive list of local providers for you to choose from. Your primary care physician may/may not be on the panel.

Courtesy of Tualatin Valley Water District

Figure A-10 Early return to work program

E. You may be asked, based on the details of the incident, to submit to a drug and alcohol test. (See the Substance Abuse Policy in the Employee Manual).

F. If your injury results in the need for immediate emergency medical care, you will remain on regular time until you are admitted to a medical care facility for medical treatment or the treating physician determines you cannot return to regular duty.

G. If you are required to submit a drug and alcohol test, you will be placed on light-duty until the Utility receives the results. During this period, you will remain on regular time.

H. If, as a result of a work-related injury, you are admitted to a medical care facility for medical treatment and the physician or other licensed health care professional determines you cannot return to regular duty or recommends followup visits, you may use sick time, vacation or comp time. If the claim is accepted by the Utility's Workers' Compensation Insurer, the insurer may reimburse your time-lost wages. In this case, the Utility will work with you to exchange the time you have taken for the wages the Utility has paid you.

I. While performing transitional work because of an accepted workers' compensation claim, you will receive your regular rate of pay for the hours worked. If you work your regular full shift, there will be no loss of wages. If you work less than your regular full shift, a temporary partial benefit will be payable in accordance with the state's Workers' Compensation Law which provides 2/3 of your 52-week average weekly wage, tax free. Any time that is not paid by workers' compensation can be supplemented with sick time, vacation or comp-time.

J. Your manager may allow you to use regular straight time, not to exceed 8 hours, when time loss is authorized for doctor visits or follow-up medical treatment. After the initial 8 hours of time loss, you may use sick time, vacation, or comp time.

K. The Utility does not determine an employee's ability to return to work. All lost time or restricted duty must be recommended by a physician or other licensed health care professional.

III. Non-Work Related Injury/Illness

A. If you sustain a non-work related injury or illness and you cannot report to work, you must notify your supervisor. Your supervisor will notify the Safety Coordinator. If you are being treated by a physician, the Safety Coordinator will send you all appropriate forms to be forwarded to your treating physician.

Courtesy of Tualatin Valley Water District

Figure A-10 Early return to work program (*continued*)

B. You must also contact Human Resources to determine if you are eligible for Family Medical Leave or Short-term Disability.

C. It is your responsibility to make sure all forms are received by your treating physician and the appropriate forms are returned to the Utility in a timely manner.

D. You must use a physician of your choosing. The Utility does not offer managed care for non-work related injury or illness. If questions regarding your ability to return to work arise, the Utility may select a treating physician for a second opinion at the Utility's expense. There is no drug and alcohol testing requirement.

E. While performing transitional work, you will receive your regular rate of pay for the hours you work. If you work a regular shift, there will be no loss in wages. If you work a partial shift and have been on short-term disability, the Utility will pay you for your time worked and you will get a partial short-term disability benefit for your time off. If you have not been on short-term disability, you will get paid for time worked; the remainder of time off can be filled in with sick time, vacation or comp time.

IV. Return to Work

A. Restricted Release

1. If your doctor determines that you are not able to return to your regular duties but are capable of some level of work activity, your doctor must complete a release-to-work form indicating the restrictions and conditions for transitional work. The Utility will attempt to provide a temporary transitional work position until you are able to resume regular duties. If the release-to-work form does not adequately describe your restrictions so that a transitional work position can be developed, the Utility may ask for more detailed information before providing a temporary transitional work position.

2. When you return your release-to-work form, the Utility will develop a transitional tasks description. The transitional duties will be discussed with you and you will be asked to sign the description of duties to acknowledge your acceptance. In order to return to work, you must sign this transitional task description.

3. All transitional work positions are subject to Utility management's review and approval. The determination may be made by any of the following: your department manager, supervisor, the Safety Coordinator, the Human Resources Director, or the Chief Executive Officer. If the workload of another department is involved, the determination will also include that manager and supervisor.

Courtesy of Tualatin Valley Water District

Figure A-10 Early return to work program (*continued*)

4. Once an approved transitional tasks description is signed, your supervisor will ensure that you are not asked to do anything outside of the duties described in the transitional tasks description. It is your responsibility to follow all restrictions set by your physician. It is also your responsibility to notify the Utility if you are asked to perform duties outside the restrictions or the job is going beyond your physical capabilities.
5. While performing transitional work, you will still be expected to follow all Utility policies.
6. This is a transitional program in which the nature of tasks may increase in level of physical demands as you recover. You must report any change in restrictions within one business day to your supervisor. This program is not intended to create transitional work positions.
7. The transitional duties will be for the length of time your doctor requests, but not for more than 60 days. After 60 days, the work availability and restrictions will be re-evaluated on a case-by-case basis.
8. If transitional work positions are limited, a worker with an active Workers' Compensation Claim, depending on the restrictions, will take priority.

V. Return to Full Duty

A. To return to full and unrestricted duty, you are required to provide a release from your treating physician. An additional release by the Utility's physician of choice may be required.

The Utility may modify, change, or discontinue the Early Return to Work Program or conditions of the program at any time dependent on Utility needs.

Courtesy of Tualatin Valley Water District

Figure A-10 Early return to work program (*continued*)

Hazard Summary/Action List

Date	Source	Site/ Equipment	Hazard	Safety Committee Review	Action Leader(s)	Action Items	Completion Date (Goal)	Completion Date (Actual)	Status	Further Explaination
01/11/11	Injury Report	Parking lot	Employee fell on black ice when walking across the parking lot.	Yes	Mike J, Collin F	Look into the availability of ice/frost indicators that will notify employees of potentially slippery conditions	02/11/11	02/11/11	Done	Collin has purchased indicators and will be installing them by the end of February.

Courtesy of Tualatin Valley Water District

Figure A-11 Hazard summary/action list

HOT WORK PERMIT		File #
Contractor:	Issued Date:	Issued Time:
Work Area:		Valid until:
Job Description:		

Items To Check	Yes	No	Remarks
Area			
1. Has the area been roped off?	☐	☐	
2. Have flammable materials been removed?	☐	☐	
3. If flammables cannot be removed, have they been protected from spark contact?	☐	☐	
4. Has the area been checked for gases?	☐	☐	
5. Is wind direction a factor?	☐	☐	
6. Is a Fire Watch required?	☐	☐	
7. Are all openings in floors and walls within 10 m (35 ft) covered to prevent hot sparks from falling below or entering walls?	☐	☐	
Equipment			
1. Has equipment been:			
Isolated?	☐	☐	
Depressured or drained?	☐	☐	
Steamed or washed?	☐	☐	
Blanked at all process connections?	☐	☐	
Tested for gases?	☐	☐	
2. Is equipment under pressure?	☐	☐	
What pressure?			
Chemicals involved? Specify.	☐	☐	
Above or below liquid level?	☐	☐	
General			
1. Has a fire extinguisher been provided?	☐	☐	
2. Are safe job procedures understood by all?	☐	☐	
3. Are fire resistant curtains / UV shields required?	☐	☐	
4. Are there smoke detectors, heat detectors or sprinkler heads that could be affected by this work?	☐	☐	
5. Are there other safety considerations not discussed above? Specify.	☐	☐	

Issued by:	Date: Time:	Comments
Checked by:	Date: Time:	
Job Completed:	Date: Time:	

All questions must be answered. If not applicable, use N/A.

Area to be checked 3 hours after completion of job to ensure no fire:		
Checked by:	Date: Time:	Comments:

Courtesy of City of Pompano Beach, Florida

Figure A-12 Hot work permit

Instructions for Completing a Job Safety Analysis Form

Job Safety Analysis (JSA) is an important analyzing tool that works by finding hazards and eliminating or minimizing them before the job is performed, and before hazards have a chance to cause injuries or do damage. Use JSA for job clarification and hazard awareness, as a guide in new employee training, for periodic contact with and retraining of senior employees, as a refresher on jobs that run infrequently, and for informing employees of specific job hazards and protective measures. JSA can also be used as an incident investigation tool.

Set priorities for doing JSAs: jobs that have a history of causing injury or damage, jobs that have produced disabling injuries, jobs with high potential for disabling injury or death, and new jobs. Select a job to be analyzed. Before filling out a JSA form, consider the following: The purpose of the job—What has to be done? Who has to do it? The activities involved—How is it done? When is it done? Where is it done? In summary, to complete the JSA form, consider the purpose of the job, the activities it involves, and the hazards it presents. If unfamiliar with a particular job or operation, interview an employee who is. In addition, observing an employee performing the job, or walking through the operation step-by-step may give additional insight into potential hazards. Consider videotaping the job and analyzing the tape. Here's how to conduct the three parts of a JSA.

Sequence of Basic Job Steps

Examining a specific job by breaking it down into a series of steps or tasks enables discovery of potential hazards employees may encounter. Each job or operation consists of a set of steps or tasks. For example, the job might be to move a box from a conveyor in the receiving area to a shelf in the storage area. To determine where a step begins or ends, look for a change of activity, direction, or movement. Picking up the box from the conveyor and placing it on a handtruck is one step. The next step might be to push the loaded handtruck to the storage area (a change in activity). Moving the boxes from the truck and placing them on the shelf is another step. The final step might be returning the handtruck to the receiving area. Be sure to list all the steps needed to perform the job. Some steps may not be performed each time; an example could be checking the casters on the handtruck. However, if that step is generally part of the job, it should be listed.

Potential Hazards

A hazard is a potential danger. The purpose of the JSA is to identify ALL hazards—both those produced by the environment or conditions and those connected with the job procedure. Examine each step carefully to find and identify hazards—the actions, conditions, and possibilities that could lead to injury, illness, or damage. Consider the following hazard types:

Chemical Hazards
__Inhalation
__Skin contact
__Absorption
__Injection
__Ingestion

Biological Hazards
__Bloodborne Pathogens
__Brucellosis
__Building-Related Illness (BRI)
__Legionnaires' Disease
__Mold
__Plant and Insect Poisons
__Tuberculosis (TB)
__Water and Wastewater

Physical Hazards
__Electrical
__Fire/Explosion
__Noise
__Radiation
__Thermal Stress
__Caught In/On/Between; Pinch Points
__Slips/Falls
__Striking Against
__Struck By

Ergonomic Hazards
__Repetition
__Forceful Exertions
__Awkward Postures
__Contact Stress
__Vibration
__Work Area Design

Recommended Action or Procedure

After evaluating and recording the job steps and potential hazards, decide what actions or procedures are necessary to eliminate or minimize those hazards that could lead to an injury, illness, or damage. Begin by trying to: 1) engineer the hazard out; 2) provide guards, safety devices, etc.; 3) provide personal protective equipment (PPE); 4) provide job instruction training; 5) maintain good housekeeping; 6) ensure good ergonomics (positioning the person in relation to the machine or other elements in such a way as to improve safety).

List the recommended safe operating procedures. Begin with an action word. Say exactly what needs to be done to correct the hazard, such as, "Lift using your leg muscles." Avoid general statements such as "be careful." List the required or recommended PPE necessary to perform each step of the job. Provide a recommended action or procedure for each hazard. Serious hazards should be corrected immediately. The JSA should then be changed to reflect the new conditions. Finally, review all the input for accuracy and completeness. Determine if the recommended actions or procedures have been put in place. Reevaluate the job safety analysis as necessary.

Job Safety Analysis (JSA) Template	Effective Date:
	Authority:

Job or Operation Title:		
Division/Department:		
Job Location:		
Title of employee(s) performing job:		
Date JSA conducted:		
JSA conducted by:		
JSA verified by:		
Special or primary hazards:		
PPE required for the job:		
Basic Job Steps	**Existing/Potential Hazards**	**Hazard Controls & Safe Work Methods**

Page 1 of 1

Figure A-13 Job safety analysis

Job Safety Analysis

Process:

TASK[1]	HAZARD[2]	CAUSE[3]	PREVENTION[4]
Emergency Pump Station Disassembly	Falls	Use of ladder; working on trailer	Review proper use/positioning
	Suspended loads	Lifting pipe sections with excavator	Taglines; inspect rigging; review lift capacity/weights; do not work under load
	Bodily injury	Working with and around tools and equipment.	PPE (see below)
	Slips and trips	Slippery surfaces rain/water), tight work area	PPE (see below); organize work; maintain good housekeeping
	Ergonomic	Lifting; tight work area, working overhead; Repetitive motion	Organize work; use proper lifting techniques; work on ladders; take occasional breaks
	Caught between/struck-by	Working around equipment; setting pipe sections on trailer	Restrict working between un stabile pipe sections and around excavator.
	Noise	Generator; impact wrench	PPE (see below)
	Electrical	Generator; impact wrench	GFCI

[1] Typical tasks: General use, routine maintenance, routine operation, shutdown unit,
[2] Typical hazards: Inhalation, eye injury, trauma, foot injury, hand injury, head injury, fire, electrical shock, fall
[3] Activity that causes the hazard
[4] Actions that can be used to minimize hazard

PPE:

R/AN/NR	PPE
x	General work gloves
	Chemical resistant gloves
	Voltage rated gloves
	Nitrile Gloves
	Vibration Gloves
x	Safety boots
	Safety rain boots
x	Safety Glasses
	Face Shield
	Splash googles
	N95 dust mask
x	Raingear
	Hooded Tyvek cover-alls

R/AN/NR	PPE
	Respirator
x	Hardhat
x	Class II high visibility vest/T-shirt
	Class II high visibility vest **only**
x	Ear plugs or muffs
	Ear plugs **and** muffs
	Chaps
	WZTC
	Shoring
	Confined space entry equipment
	Amp meter
	Fall Protection
	Apron

PROCEDURE:

1. Stretch prior to physical activity to warm up muscles.
2. Inspect tools and equipment prior use
3. Inspect and test GFCI and cords.
4. Use ladders properly
   - Read and follow all labels/markings on the ladder.
   - Always inspect the ladder prior to using it. If the ladder is damaged, it must be removed from service and tagged until repaired or discarded.
   - Always maintain a 3-point (two hands and a foot, or two feet and a hand) contact on the ladder when climbing. Keep your body near the middle of the step and always face the ladder while climbing (see diagram).
   - Ladders must be free of any slippery material on the rungs, steps or feet.
   - Do not use a self-supporting ladder (e.g., step ladder) as a single ladder or in a partially closed position.

Figure A-14 Sample job safety analysis

Job Safety Analysis

▪ Do not use the top step/rung of a ladder as a step/rung unless it was designed for that purpose. ▪ Use a ladder only on a stable and level surface, unless it has been secured (top or bottom) to prevent displacement. ▪ Do not place a ladder on boxes, barrels or other unstable bases to obtain additional height. ▪ Do not move or shift a ladder while a person or equipment is on the ladder. ▪ An extension or straight ladder used to access an elevated surface must extend at least 3 feet above the point of support (see diagram). Do not stand on the three top rungs of a straight, single or extension ladder. ▪ The proper angle for setting up a ladder is to place its base a quarter of the working length of the ladder from the wall or other vertical surface (see diagram). ▪ A ladder placed in any location where it can be displaced by other work activities must be secured to prevent displacement or a barricade must be erected to keep traffic away from the ladder. ▪ Be sure that all locks on an extension ladder are properly engaged. ▪ Do not exceed the maximum load rating of a ladder. Be aware of the ladder's load rating and of the weight it is supporting, including the weight of any tools or equipment.
5. Take occasional breaks to avoid repetitive motion injury
6. Ensure firm footing during operations.
7. Operate equipment in strict accordance with Manufacturer's instructions.
8. Hearing Protection must be worn when running the generator, impact wrench.
9. Report any observed defect or safety hazard to your supervisor immediately.
10. Report any injuries immediately.

Figure A-14 Sample job safety analysis (*continued*)

AWWA M3 SAFETY MANAGEMENT FOR UTILITIES

Passing Criteria
Insert Water Utility Name
Valve, Maintenance, & Construction Crew

PCT Tests	Passing Criteria
Step Test	3 Levels – 3 minutes each (72, 104, 124)
Twist Test	Rotate entire body to the right and left each way. 5×
Maximal Lift, Carry & Flexibility Test	Lift 50-lb bag off the floor, carry 50 ft, and set on floor. Squat onto your heels, hold for 3 seconds, then stand up on your tiptoes, arms stretched overhead and hold for 3 seconds. 3× Lift 75-lb 4 in. valve off of the floor, carry 50 ft, and set back on the floor. Lift 90-lb crate off the floor to 43 in. high top landing on stairs and set back on the floor. 2× Lift 30-lb crate off the floor to 67 in. high and set back on the floor. 3×
Push/Pull Tests	95-lb vertical pull upwards hold 3 sec. 2× 80-lb 40-in. horizontal push 3 sec. 3× 100-lb 40-in. horizontal pull 3 sec. 3× Pull 125-lb drag bag 40 ft.
Shovel Test	Scoop 13.6-lb shovel from the floor to 45 in. and back to the floor. 2×
Double Step Test	Step up onto second and fourth rung of stepladder and back down. 3×
Valve Turn Test	
Body Mechanics (Lifting)	Erect spine (3), load close to body (2), pivot feet (2), bend knees (2), wide base of support (1). Total of 10 points

Courtesy of Tualatin Valley Water District

Figure A-15 Passing criteria

Enter Utility Name

Physical Capacities Requirements

Position: Click here to enter text.

In a 9 hour workday, the worker will be:

	(Total hours at one time)		*(Total hours during entire workday)*
Standing	1/2	**Standing**	1/2
Sitting	1/2	**Sitting**	1/2

Job requires repetitive use of hands for:

	Right	*Left*	*Both*
Simple Grasping	☐	☐	☐
Pushing & Pulling	☐	☐	☐
Fine Manipulating	☐	☐	☐

Job requires repetitive movement of feet as in operating foot controls:

Right ☐ *Left* ☐ *Both* ☐

"Occasionally" equals 1% to 33% *"Frequently" equals 34% to 64%* *"Continually" equals 65% to 100%*

Job requires worker to lift:	**Never**	**Occasionally**	**Frequently**	**Continually**
Up to 5 lbs.	☐	☐	☐	☐
6 to 10 lbs.	☐	☐	☐	☐
11 to 20 lbs.	☐	☐	☐	☐
21 to 50 lbs.	☐	☐	☐	☐
51 to 100 lbs.	☐	☐	☐	☐
Job requires worker to carry:	**Never**	**Occasionally**	**Frequently**	**Continually**
Up to 5 lbs.	☐	☐	☐	☐
6 to 10 lbs.	☐	☐	☐	☐
11 to 20 lbs.	☐	☐	☐	☐
21 to 50 lbs.	☐	☐	☐	☐
51 to 100 lbs.	☐	☐	☐	☐
Job requires worker to:	**Never**	**Occasionally**	**Frequently**	**Continually**
Bend	☐	☐	☐	☐
Squat	☐	☐	☐	☐
Crawl	☐	☐	☐	☐
Reach above shoulder level	☐	☐	☐	☐
Walk or work on uneven surfaces	☐	☐	☐	☐
Job requires these activities:	**Never**	**Occasionally**	**Frequently**	**Continually**
Being around machinery	☐	☐	☐	☐
Operating heavy equipment	☐	☐	☐	☐
Exposure to dust, fumes & gases	☐	☐	☐	☐

Comments: Click here to enter text.

Manager's Signature:______________________________**Date:**____________________

Adapted from Tualatin Valley Water District with permission

Figure A-16 Physical capacities requirements

Return form to:

AWWA M3 SAFETY MANAGEMENT FOR UTILITIES

RELEASE TO RETURN TO WORK

Name of worker	Claim number

Please fill out this form and return it to us at the address indicated above.

1. Is the worker medically stationary? ☐ Yes ☐ No Date ____________ (Provide closing information and complete Form 827.)
   Next scheduled appointment date ____________
2. Worker is released to:
   ☐ full duty without limitations Date ____________ (Do not complete lines 3 through 11. Sign below.)
   ☐ modified duty from (date) ________ through (date) ____________ (specify limitations below.)
   ☐ modified hours — specify ____________ from (date) ________ through (date) ________

Hours:	No limitations	1	2	3	4	5	6	7	8
3. In an eight-hour workday, worker can stand/walk a total of	☐	☐	☐	☐	☐	☐	☐	☐	☐
4. At one time, worker can stand/walk	☐	☐	☐	☐	☐	☐	☐	☐	☐
5. In an eight-hour workday, worker can sit a total of	☐	☐	☐	☐	☐	☐	☐	☐	☐
6. At one time, worker can sit	☐	☐	☐	☐	☐	☐	☐	☐	☐

7. The worker is released to return to work in the following range for lifting, carrying, pushing/pulling:

Pounds	<10	10	15	20	25	30	35	40	45	50	55	60	65	70	75	80	85	90	95	100	>100
Occasionally	☐	☐	☐	☐	☐	☐	☐	☐	☐	☐	☐	☐	☐	☐	☐	☐	☐	☐	☐	☐	☐
Frequently	☐	☐	☐	☐	☐	☐	☐	☐	☐	☐	☐	☐	☐	☐	☐	☐	☐	☐	☐	☐	☐

8. Worker can use hands for repetitive:	**Right**	**Left**	
a. Fine manipulation	☐ Yes ☐ No	☐ Yes ☐ No	Dominant hand
b. Pushing and pulling	☐ Yes ☐ No	☐ Yes ☐ No	☐ Right ☐ Left
c. Simple grasping	☐ Yes ☐ No	☐ Yes ☐ No	
d. Keyboarding	☐ Yes ☐ No	☐ Yes ☐ No	

9. Worker can use feet for repetitive raising and pushing (as in operating foot controls): ☐ Yes ☐ No

10. Worker is able to:	**Continuous** 67-100% of the day	**Frequently** 34-66% of the day	**Occasionally** 6-33% of the day	**Intermittently** 1-5% of the day	**Not at all**
a. Stoop/bend	☐	☐	☐	☐	☐
b. Crouch	☐	☐	☐	☐	☐
c. Crawl	☐	☐	☐	☐	☐
d. Kneel	☐	☐	☐	☐	☐
e. Twist	☐	☐	☐	☐	☐
f. Climb	☐	☐	☐	☐	☐
g. Balance	☐	☐	☐	☐	☐
h. Reach	☐	☐	☐	☐	☐
i. Push/pull	☐	☐	☐	☐	☐

11. Other functional limitations or modifications necessary in worker's employment:

Additional comments may be written on back of form.

Signature of medical service provider*	Printed name	Date

Adapted from Tualatin Valley Water District with permission

Figure A-17 Release to return to work

AWWA M3 SAFETY MANAGEMENT FOR UTILITIES

Pre-Placement Physical Capacities Test

Insert Utility Name

Applicant Name: ____________________

I. STEP TEST

Stages	Metronome	Time Stopped	Comments
Level 1	72		
Level 2	104		
Level 3	124		

Must complete 3 minutes at each level to pass.

Pass/Fail

II. TWIST TEST: Rotate entire body to the right and left each way 5×. ________

III. MAXIMUM LIFT, CARRY, & FLEXIBILITY TEST

A. Lift 50-lb bag off the floor, carry 50 ft, and set on floor. ________
B. Squat onto your heels hold for 3 seconds, then stand up on tiptoes, arms stretched overhead, and hold for 3 seconds. 3× ________
C. Lift 75-lb 4-in. valve off the floor, carry 50 ft, and set back on the floor. ________
D. Lift 90 lb off the floor to 43 in. high top landing on stairs, set back on the floor. 2× ________
E. Lift 30 lb off the floor to 67 in. high and set back on the floor. 3× ________

IV. PUSH AND PULL TEST

	Distance To Floor	Time Held	Max.Weight	Limit	Reason For Termination
Vertical pull upwards				95 lb	
3 sec × 2 reps					
40 in. horizontal push				80 lb	
3 sec × 3 reps					
40 in. horizontal pull				100 lb	
3 sec × 3 reps					
Pull 125 lb drag bag					
40 ft × 1 rep					

V. SHOVEL TEST

Scoop 13.6 lb shovel from floor to 45 in. and back to floor. 25× ________

VI. DOUBLE CLIMB TEST

Step up onto second and fourth rung of stepladder and back down. 3× ________

VII. VALVE TURN TEST:

VIII. BODY MECHANICS:

Erect spine (3), load close to body (2), pivot feet (2), bend knees (2), wide base of support (1). Total of 10 points.

________ (0-10 points)

Comments:

____
____
____
____
____

Evaluator's Name____________________ Date__________

Figure A-18 Pre-placement physical capacities test

OSHA QUICK CARD™

Hazard Communication Safety Data Sheets

The Hazard Communication Standard (HCS) requires chemical manufacturers, distributors, or importers to provide Safety Data Sheets (SDSs) (formerly known as Material Safety Data Sheets or MSDSs) to communicate the hazards of hazardous chemical products. As of June 1, 2015, the HCS will require new SDSs to be in a uniform format, and include the section numbers, the headings, and associated information under the headings below:

Section 1, Identification includes product identifier; manufacturer or distributor name, address, phone number; emergency phone number; recommended use; restrictions on use.

Section 2, Hazard(s) identification includes all hazards regarding the chemical; required label elements.

Section 3, Composition/information on ingredients includes information on chemical ingredients; trade secret claims.

Section 4, First-aid measures includes important symptoms/ effects, acute, delayed; required treatment.

Section 5, Fire-fighting measures lists suitable extinguishing techniques, equipment; chemical hazards from fire.

Section 6, Accidental release measures lists emergency procedures; protective equipment; proper methods of containment and cleanup.

Section 7, Handling and storage lists precautions for safe handling and storage, including incompatibilities.

Section 8, Exposure controls/personal protection lists OSHA's Permissible Exposure Limits (PELs); Threshold Limit Values (TLVs); appropriate engineering controls; personal protective equipment (PPE).

Section 9, Physical and chemical properties lists the chemical's characteristics.

Section 10, Stability and reactivity lists chemical stability and possibility of hazardous reactions.

Section 11, Toxicological information includes routes of exposure; related symptoms, acute and chronic effects; numerical measures of toxicity.

Section 12, Ecological information*

Section 13, Disposal considerations*

Section 14, Transport information*

Section 15, Regulatory information*

Section 16, Other information, includes the date of preparation or last revision.

*Note: Since other Agencies regulate this information, OSHA will not be enforcing Sections 12 through 15 (29 CFR 1910.1200(g)(2)).

Employers must ensure that SDSs are readily accessible to employees.

See Appendix D of OSHA 1910.1200 for a detailed description of SDS contents.

For more information: www.osha.gov, (800) 321-OSHA (6742)

From Occupational Safety and Health Administration, US Department of Labor
www.osha.gov/Publications/HazComm_QuickCard_SafetyData.html

Figure A-19 OSHA Quick Card

SAMPLE SAFETY DATA SHEET

Issuing Date January 5, 2015 **Revision Date** June 12, 2015 **Revision Number** 1

1. IDENTIFICATION OF THE SUBSTANCE/PREPARATION AND OF THE COMPANY/UNDERTAKING

Product identifier

Product Name **XXXXX Regular-Bleach₁**

Other means of identification

EPA Registration Number 5813-100

Recommended use of the chemical and restrictions on use

Recommended use Household disinfecting, sanitizing, and laundry bleach

Uses advised against No information available

Details of the supplier of the safety data sheet

Supplier Address
The XXXXX Company
1221 Broadway
Oakland, CA 94612

Phone: 1-510-XXX-XXXX

Emergency telephone number

Emergency Phone Numbers For Medical Emergencies, call: 1-800-446-1014
For Transportation Emergencies, call Chemtrec: 1-800-424-9300

Page 1 / 10

Figure A-20 Safety data sheet sample

XXXXX Regular-Bleach₁ **Revision Date** June 12, 2015

2. HAZARDS IDENTIFICATION

Classification

This chemical is considered hazardous by the 2012 OSHA Hazard Communication Standard (29 CFR 1910.1200).

Skin corrosion/irritation	Category 1
Serious eye damage/eye irritation	Category 1

GHS Label elements, including precautionary statements

Emergency Overview

Signal word **Danger**

Hazard Statements
Causes severe skin burns and eye damage
Causes serious eye damage

Appearance Clear, pale yellow **Physical State** Thin liquid **Odor** Bleach

Precautionary Statements - Prevention
Wash face, hands and any exposed skin thoroughly after handling.
Wear protective gloves, protective clothing, face protection, and eye protection such as safety glasses.

Precautionary Statements - Response
Immediately call a poison center or doctor.
If swallowed: Rinse mouth. Do NOT induce vomiting.
If on skin (or hair): Take off immediately all contaminated clothing. Rinse skin with water.
Wash contaminated clothing before reuse.
If inhaled: Remove person to fresh air and keep comfortable for breathing.
Specific treatment (see supplemental first aid instructions on this label).
If in eyes: Rinse cautiously with water for several minutes. Remove contact lenses, if present and easy to do. Continue rinsing.

Precautionary Statements - Storage
Store locked up.

Precautionary Statements - Disposal
Dispose of contents in accordance with all applicable federal, state, and local regulations.

Hazards not otherwise classified (HNOC)
Although not expected, heart conditions or chronic respiratory problems such as asthma, chronic bronchitis, or obstructive lung disease may be aggravated by exposure to high concentrations of vapor or mist.

Product contains a strong oxidizer. Always flush drains before and after use.

Page 2 / 10

Figure A-20 Safety data sheet sample (*continued*)

XXXXX Regular-Bleach₁ **Revision Date** June 12, 2015

Unknown Toxicity
Not applicable.

Other information
Very toxic to aquatic life with long lasting effects.

Interactions with Other Chemicals
Reacts with other household chemicals such as toilet bowl cleaners, rust removers, acids, or products containing ammonia to produce hazardous irritating gases, such as chlorine and other chlorinated compounds.

Chemical Name	CAS-No	Weight %	Trade Secret
Sodium hypochlorite	7681-52-9	5 - 10	*

* The exact percentage (concentration) of composition has been withheld as a trade secret.

4. FIRST AID MEASURES

First aid measures

General Advice Call a poison control center or doctor immediately for treatment advice. Show this safety data sheet to the doctor in attendance.

Eye Contact Hold eye open and rinse slowly and gently with water for 15 - 20 minutes. Remove contact lenses, if present, after the first 5 minutes, then continue rinsing eye. Call a poison control center or doctor for treatment advice.

Skin Contact Take off contaminated clothing. Rinse skin immediately with plenty of water for 15-20 minutes. Call a poison control center or doctor for treatment advice.

Inhalation Move to fresh air. If breathing is affected, call a doctor.

Ingestion Have person sip a glassful of water if able to swallow. Do not induce vomiting unless told to do so by a poison control center or doctor. Do not give anything by mouth to an unconscious person. Call a poison control center or doctor immediately for treatment advice.

Protection of First-aiders Avoid contact with skin, eyes, and clothing. Use personal protective equipment as required. Wear personal protective clothing (see section 8).

Most important symptoms and effects, both acute and delayed

Most Important Symptoms and Effects Burning of eyes and skin.

Indication of any immediate medical attention and special treatment needed

Notes to Physician Treat symptomatically. Probable mucosal damage may contraindicate the use of gastric lavage.

Page 3 / 10

Figure A-20 Safety data sheet sample (*continued*)

XXXXX Regular-Bleach$_1$ **Revision Date** June 12, 2015

5. FIRE-FIGHTING MEASURES

Suitable Extinguishing Media
Use extinguishing measures that are appropriate to local circumstances and the surrounding environment.

Unsuitable Extinguishing Media
CAUTION: Use of water spray when fighting fire may be inefficient.

Specific Hazards Arising from the Chemical
This product causes burns to eyes, skin, and mucous membranes. Thermal decomposition can release sodium chlorate and irritating gases and vapors.

Explosion Data

Sensitivity to Mechanical Impact None.

Sensitivity to Static Discharge None.

Protective equipment and precautions for firefighters
As in any fire, wear self-contained breathing apparatus pressure-demand, MSHA/NIOSH (approved or equivalent) and full protective gear.

6. ACCIDENTAL RELEASE MEASURES

Personal precautions, protective equipment and emergency procedures

Personal Precautions Avoid contact with eyes, skin, and clothing. Ensure adequate ventilation. Use personal protective equipment as required. For spills of multiple products, responders should evaluate the MSDSs of the products for incompatibility with sodium hypochlorite. Breathing protection should be worn in enclosed and/or poorly-ventilated areas until hazard assessment is complete.

Other Information Refer to protective measures listed in Sections 7 and 8.

Environmental precautions

Environmental Precautions This product is toxic to fish, aquatic invertebrates, oysters, and shrimp. Do not allow product to enter storm drains, lakes, or streams. See Section 12 for ecological Information.

Methods and material for containment and cleaning up

Methods for Containment Prevent further leakage or spillage if safe to do so.

Methods for Cleaning Up Absorb and containerize. Wash residual down to sanitary sewer. Contact the sanitary treatment facility in advance to assure ability to process washed-down material.

Page 4 / 10

Figure A-20 Safety data sheet sample (*continued*)

XXXXX Regular-Bleach₁ **Revision Date** June 12, 2015

Precautions for safe handling

Handling Handle in accordance with good industrial hygiene and safety practice. Avoid contact with skin, eyes, and clothing. Do not eat, drink, or smoke when using this product.

Conditions for safe storage, including any incompatibilities

Storage Store away from children. Reclose cap tightly after each use. Store this product upright in a cool, dry area, away from direct sunlight and heat to avoid deterioration. Do not contaminate food or feed by storage of this product.

Incompatible Products Toilet bowl cleaners, rust removers, acids, and products containing ammonia.

8. EXPOSURE CONTROLS/PERSONAL PROTECTION

Control parameters

Exposure Guidelines

Chemical Name	ACGIH TLV	OSHA PEL	NIOSH IDLH
Sodium hypochlorite 7681-52-9	None	None	None

ACGIH TLV: American Conference of Governmental Industrial Hygienists - Threshold Limit Value. OSHA PEL: Occupational Safety and Health Administration - Permissible Exposure Limits. NIOSH IDLH: Immediately Dangerous to Life or Health.

Appropriate engineering controls

Engineering Measures Showers
Eyewash stations
Ventilation systems

Individual protection measures, such as personal protective equipment

Eye/Face Protection If splashes are likely to occur: Wear safety glasses with side shields (or goggles) or face shield.

Skin and Body Protection Wear rubber or neoprene gloves and protective clothing such as long-sleeved shirt.

Respiratory Protection If irritation is experienced, NIOSH/MSHA approved respiratory protection should be worn. Positive-pressure supplied air respirators may be required for high airborne contaminant concentrations. Respiratory protection must be provided in accordance with current local regulations.

Hygiene Measures Handle in accordance with good industrial hygiene and safety practice. Wash hands after direct contact. Do not wear product-contaminated clothing for prolonged periods. Remove and wash contaminated clothing before re-use. Do not eat, drink, or smoke when using this product.

Page 5 / 10

Figure A-20 Safety data sheet sample (*continued*)

9. PHYSICAL AND CHEMICAL PROPERTIES

Physical and Chemical Properties

Physical State	Thin liquid		
Appearance	Clear	**Odor**	Bleach
Color	Pale yellow	**Odor Threshold**	No information available

Property	**Values**	**Remarks/ Method**
pH	~12	None known
Melting/freezing point	No data available	None known
Boiling point / boiling range	No data available	None known
Flash Point	Not flammable	None known
Evaporation rate	No data available	None known
Flammability (solid, gas)	No data available	None known
Flammability Limits in Air		
Upper flammability limit	No data available	None known
Lower flammability limit	No data available	None known
Vapor pressure	No data available	None known
Vapor density	No data available	None known
Specific Gravity	~1.1	None known
Water Solubility	Soluble	None known
Solubility in other solvents	No data available	None known
Partition coefficient: n-octanol/water	No data available	None known
Autoignition temperature	No data available	None known
Decomposition temperature	No data available	None known
Kinematic viscosity	No data available	None known
Dynamic viscosity	No data available	None known
Explosive Properties	Not explosive	
Oxidizing Properties	No data available	

Other Information

Softening Point	No data available
VOC Content (%)	No data available
Particle Size	No data available
Particle Size Distribution	No data available

10. STABILITY AND REACTIVITY

Reactivity
Reacts with other household chemicals such as toilet bowl cleaners, rust removers, acids, or products containing ammonia to produce hazardous irritating gases, such as chlorine and other chlorinated compounds.

Chemical stability
Stable under recommended storage conditions.

Possibility of Hazardous Reactions
None under normal processing.

Conditions to avoid
None known based on information supplied.

Incompatible materials
Toilet bowl cleaners, rust removers, acids, and products containing ammonia.

Hazardous Decomposition Products
None known based on information supplied.

Figure A-20 Safety data sheet sample (*continued*)

XXXXX Regular-Bleach[1] **Revision Date** June 12, 2015

11. TOXICOLOGICAL INFORMATION

Information on likely routes of exposure

Product Information .

Inhalation Exposure to vapor or mist may irritate respiratory tract and cause coughing. Inhalation of high concentrations may cause pulmonary edema.

Eye Contact Corrosive. May cause severe damage to eyes.

Skin Contact May cause severe irritation to skin. Prolonged contact may cause burns to skin.

Ingestion Ingestion may cause burns to gastrointestinal tract and respiratory tract, nausea, vomiting, and diarrhea.

Component Information

Chemical Name	LD50 Oral	LD50 Dermal	LC50 Inhalation
Sodium hypochlorite 7681-52-9	8200 mg/kg (Rat)	>10000 mg/kg (Rabbit)	-

Information on toxicological effects

Symptoms May cause redness and tearing of the eyes. May cause burns to eyes. May cause redness or burns to skin. Inhalation may cause coughing.

Delayed and immediate effects as well as chronic effects from short and long-term exposure

Sensitization No information available.

Mutagenic Effects No information available.

Carcinogenicity The table below indicates whether each agency has listed any ingredient as a carcinogen.

Chemical Name	ACGIH	IARC	NTP	OSHA
Sodium hypochlorite 7681-52-9	-	Group 3	-	-

IARC (International Agency for Research on Cancer)
Group 3 - Not Classifiable as to Carcinogenicity in Humans

Reproductive Toxicity No information available.

STOT - single exposure No information available.

STOT - repeated exposure No information available.
Chronic Toxicity Carcinogenic potential is unknown.
Target Organ Effects Respiratory system, eyes, skin, gastrointestinal tract (GI).

Aspiration Hazard No information available.

Page 7 / 10

Figure A-20 Safety data sheet sample (*continued*)

XXXXX Regular-Bleach₁ **Revision Date** June 12, 2015

Numerical measures of toxicity - Product Information

The following values are calculated based on chapter 3.1 of the GHS document

ATEmix (oral)
54 g/kg
ATEmix (inhalation-dust/mist)
58 mg/L

12. ECOLOGICAL INFORMATION

Ecotoxicity
Very toxic to aquatic life with long lasting effects.

This product is toxic to fish, aquatic invertebrates, oysters, and shrimp. Do not allow product to enter storm drains, lakes, or streams.

Persistence and Degradability
No information available.

Bioaccumulation
No information available.

Other adverse effects
No information available.

13. DISPOSAL CONSIDERATIONS

Disposal methods
Dispose of in accordance with all applicable federal, state, and local regulations. Do not contaminate food or feed by disposal of this product.

Contaminated Packaging
Do not reuse empty containers. Dispose of in accordance with all applicable federal, state, and local regulations.

14. TRANSPORT INFORMATION

DOT	Not restricted.
TDG	Not restricted for road or rail.
ICAO	Not restricted, as per Special Provision A197, Environmentally Hazardous Substance exception.
IATA	Not restricted, as per Special Provision A197, Environmentally Hazardous Substance exception.
IMDG/IMO	Not restricted, as per IMDG Code 2.10.2.7, Marine Pollutant exception.

Page 8 / 10

Figure A-20 Safety data sheet sample (*continued*)

15. REGULATORY INFORMATION

Chemical Inventories

TSCA All components of this product are either on the TSCA 8(b) Inventory or otherwise exempt from listing.

DSL/NDSL All components are on the DSL or NDSL.

TSCA - United States Toxic Substances Control Act Section 8(b) Inventory
DSL/NDSL - Canadian Domestic Substances List/Non-Domestic Substances List

U.S. Federal Regulations

SARA 313
Section 313 of Title III of the Superfund Amendments and Reauthorization Act of 1986 (SARA). This product does not contain any chemicals which are subject to the reporting requirements of the Act and Title 40 of the Code of Federal Regulations, Part 372

SARA 311/312 Hazard Categories

Acute Health Hazard	Yes
Chronic Health Hazard	No
Fire Hazard	No
Sudden Release of Pressure Hazard	No
Reactive Hazard	No

Clean Water Act
This product contains the following substances which are regulated pollutants pursuant to the Clean Water Act (40 CFR 122.21 and 40 CFR 122.42)

Chemical Name	CWA - Reportable Quantities	CWA - Toxic Pollutants	CWA - Priority Pollutants	CWA - Hazardous Substances
Sodium hypochlorite 7681-52-9	100 lb			X

CERCLA
This material, as supplied, contains one or more substances regulated as a hazardous substance under the Comprehensive Environmental Response Compensation and Liability Act (CERCLA) (40 CFR 302)

Chemical Name	Hazardous Substances RQs	Extremely Hazardous Substances RQs	RQ
Sodium hypochlorite 7681-52-9	100 lb	-	RQ 100 lb final RQ RQ 45.4 kg final RQ

EPA Statement
This chemical is a pesticide product registered by the Environmental Protection Agency and is subject to certain labeling requirements under federal pesticide law. These requirements differ from the classification criteria and hazard information required for safety data sheets and for workplace labels of non-pesticide chemicals. Following is the hazard information as required on the pesticide label:

> **DANGER: CORROSIVE.** Causes irreversible eye damage and skin burns. Harmful if swallowed. Do not get in eyes, on skin, or on clothing. Wear protective eyewear and rubber gloves when handling this product. Wash thoroughly with soap and water after handling and before eating, drinking, chewing gum, using tobacco, or using the restroom. Avoid breathing vapors and use only in a well-ventilated area.

Figure A-20 Safety data sheet sample (*continued*)

US State Regulations

California Proposition 65
This product does not contain any Proposition 65 chemicals.

U.S. State Right-to-Know Regulations

Chemical Name	New Jersey	Massachusetts	Pennsylvania	Rhode Island	Illinois
Sodium hypochlorite 7681-52-9	X	X	X	X	
Sodium chlorate 7775-09-9	X	X	X		

International Regulations

Canada
WHMIS Hazard Class
E - Corrosive material

16. OTHER INFORMATION

NFPA **Health Hazard** 3 **Flammability** 0 **Instability** 0 **Physical and Chemical Hazards** -

HMIS **Health Hazard** 3 **Flammability** 0 **Physical Hazard** 0 **Personal Protection** B

Prepared By Product Stewardship
23 British American Blvd.
Latham, NY 12110
1-800-572-6501

Revision Date June 12, 2015

Revision Note Revision Section 14.

Reference 1096036/164964.159

General Disclaimer
The information provided in this Safety Data Sheet is correct to the best of our knowledge, information and belief at the date of its publication. The information given is designed only as a guidance for safe handling, use, processing, storage, transportation, disposal, and release and is not to be considered a warranty or quality specification. The information relates only to the specific material designated and may not be valid for such material used in combination with any other materials or in any process, unless specified in the text.

End of Safety Data Sheet

This material was produced under grant 46F6-HT30 from the Occupational Safety and Health Administration, U.S. Department of Labor. It does not necessarily reflect the views or policies of the U.S. Department of Labor, nor does mention of trade names, commercial products, or organizations imply endorsement by the U.S. Government. Revisions were made to this material under grant number SH-31240-SH7 from the Occupational Safety and Health Administration, U.S. Department of Labor.

Figure A-20 Safety data sheet sample (*continued*)

AWWA M3 SAFETY MANAGEMENT FOR UTILITIES

Temporary Modified Duty Tasks

General Tasks Description:

The purpose of these temporary modified duty tasks is to facilitate your transition back to regular work duties.

Your modified duty assignments, which are listed below, are within the restrictions as identified by your treating physician. If you are assigned tasks which are outside of your restrictions, as identified by your treating physician, you are to report them to your supervisor, the Safety Coordinator, or the Human Resources Manager immediately.

You must report to the supervisor identified below. If this is a department other than the one you typically report to, review with the supervisor the expectations of that department. Any changes must be reported to and approved by the supervisor identified below.

These modified duty tasks are temporary and will be reevaluated in 60 days, when restrictions change, or when listed tasks are no longer available.

Report to:

Tasks:

Restrictions:

I have reviewed the above temporary modified duty tasks and feel that is in within my restrictions as identified by my treating physician. I also agree to notify my supervisor, the Safety Coordinator, or the Human Resources Manager if I am not able to perform assigned duties or my restrictions change.

Employee Name (Print):____

Employee Signature:____

Date:________________________

Initials: Supervisor_________ Manager__________ Safety___________

Courtesy of Tualatin Valley Water District

Figure A-21 Temporary modified duty tasks

[Insert Utility Name]	
Title: Field Operations -- Water Works Operator	**Job Grade: 3 during training 6 after training & successful completion of trial service**
Reporting to: Field Operations or Valve Crew Supervisor	**Department: Field Operations**

[Insert Water Utility Name] believes that each employee makes a significant contribution to our success. That contribution should not be limited by the assigned responsibilities. Therefore, this job description is designed to outline primary duties, qualifications and job scope, but not limit the incumbent nor the organization to just the work identified. It is our expectation that each employee will offer his/her services wherever and whenever necessary to ensure the success of our endeavors.

Overall Purpose of the job: To perform hands-on functions related to construction, operation, repair, and maintenance of the Utility's water system and related facilities. The position includes installation, repair, operations, maintenance and documentation of of water mains, services, hydrants and related appurtenances. The job requires the position to a wide variety of tools and equipment.

This position has been identified as safety sensitive because one or more of the following are essential functions of the position: Operating heavy equipment, operation of a commercial vehicle, setting up work zone traffic control, confined space work, trenching and excavation work, operation of a forklift, flagging, working with water treatment chemicals, driving (as a primary function of the position).

Essential Job Functions:
- Digging by hand and using hand and power tools are essential to this job function.
- Participate in proper construction, maintenance, operation, and repair of assigned water system infrastructure which may include: mains and related fittings, hydrants, valves, services, meters, regulators, vaults and other components using proper restraint, disinfection, and flushing methods per schedules, plans and/or the Supervisor's direction in a fashion consistent with the Utility Water System Standards and industry best practices.
- Operate various types of heavy equipment and tools, including excavation equipment, dump trucks, tank trucks, crane trucks, vacuums, hydraulic and air power tools, saws, compaction equipment, shoring systems and other miscellaneous types of tools.
- Properly maintain tools and equipment used by the crew.
- Assemble or organize parts, materials and tools appropriate to the task assuring they are available on the job site when needed to avoid unnecessary work delays.
- Performing traffic control in support of the crew's operations.
- Following sound environmental practices.
- Following sound safety practice.
- Acurately complete documention appropriate to assigned tasks in formats required, which may include mapping system updates, as-builts, work order documents, asset records, condition documentation, inventory and work time tracking among others.
- Participate in emergency system repairs and operations as required.
- Provide high quality workmanship in an efficient, cost conscious, and sustainable manner.
- Other duties as assigned.

Courtesy of Tualatin Valley Water District

1 of 6

Figure A-22 Job description (heavy physical requirements)

Knowledge Required:

- General knowledge of water distribution system operations.
- Knowledge of routine water system construction, maintenance, and repair methods and procedures.
- Can understand and interpret maps and project plans. Can create clear and understandable field drawings, maps, and other documents appropriate to assigned tasks.
- Know proper operating practices of light and heavy tools and equipment used by the crew.
- Know and follow safety practices, and use safety equipment appropriate to the assigned tasks.

Special Skills and Qualities:

- Able to take and act on written and verbal instructions.
- Able to effectively use a pick, shovel, pry bar, jackhammer and other heavy hand tools.
- Good customer relations and communication skills, including the ability to effectively and tactfully work with difficult and unreasonable customers.
- Ability to establish and maintain harmonious relationships with others both inside and outside of the Utility.
- Able to learn and effectively work with computer applications and electronic equipment routinely used by the crew.

Minimum requirements:

- High School Diploma or equivalent.
- Ability to take and pass the Water Distribution System Operator I examination after a maximum of two years of holding the position.
- Ability to work as a team member in a crew environment.
- A valid Oregon State Driver's License; have and maintain a safe driving record and the ability to obtain a Class A Commercial Driver's License (CDL) with airbrakes and tanker endorsement within six months of holding the position.
- Have and maintain a safe driving record.

Working Conditions:

- Work a minimum of nine hours per day and other hours as required.
- Able to work variable work schedule to accommodate project and customer constraints.
- Ability to have regular and predictable attendance.
- Take on-call duty as needed.
- Work in adverse weather conditions, and perform tasks requiring significant physical exertion, including repetitive lifting, digging, prying, and kneeling on a daily basis.
- Respond and work, without advanced notice, in times of adverse weather, natural disaster, emergency, or other unusual events as determined by the Utility.

Courtesy of Tualatin Valley Water District

Figure A-22 Job description (heavy physical requirements) (*continued*)

Maintenance and Construction Crew Waterworks Operator

PHYSICAL ACTIVITIES

In an eight-hour day, this job requires:

R – Rarely Less than 1 hr or 1–5% per day

O – Occasionally 1–2.5 hrs or 10–33% per day
10% = 4–6×/hr, 33% = 20×/hr or < 60×/day)

F – Frequently 2.5–5.5 hrs or 34–66% per day
(34% = > 20×/hr, 66%=120×/hr or < 720x/day)

C – Continually 5.5 – 8 hrs or 67–100% per day
(67% = > 120×/hr or > 720×/day)

	R	O	F	C	Notes
Assembly		X			Pipe assembly, tighten fittings
Sitting	X				Vehicles, meetings
Standing			X		Repairs, flagging, operating equip, pogo
Walking				X	Uneven ground, 100'+ in field
Squatting/crouching		X			Repair water systems, trench
Driving			X		Vehicles
Crawling	X				Trench
Kneeling		X			Repair meters
Digging/shoveling	X				Landscape clean up
Stooping/bending			X		Install/ repair pipe
Turn/twisting			X		Handle equipment, handle pipes
Balancing		X			Walking on pipe in trench
Reaching forward			X		Install/ repair lines
Reaching overhead	X				In ditch, reaching up for tools, from trench for tools
Climbing		X			16" & 26" step into dump truck, shallow ditches, truck
Ladder climbing	X				6'-10' ladder in/out of ditch
Wrist turning			X		Hand tools, wrench
Grasping				X	Tools, valves, wrench
Pinch/squeeze				X	Hand tools
Finger manipulation				X	Hand tools

Courtesy of Tualatin Valley Water District

3 of 6

Figure A-23 Physical capacities requirements (heavy): maintenance and construction crew waterworks operator

Maintenance and Construction Crew Waterworks Operator

LIFTING

In an eight-hour day, this job requires: **R - Rarely** Less than 1 hr or 1–5% per day **O - Occasionally** 1–2.5 hrs or 10–33% per day 10% = 4–6×/hr, 33% = 20×/hr or < 60×/day)				**F - Frequently** 2.5–5.5 hrs or 34–66% per day (34% = > 20×/hr, 66%=120×/hr or < 720x/day) **C - Continually** 5.5 – 8 hrs or 67–100% per day (67% = > 120×/hr or > 720×/day)			
Weight	Object	Lowest point	Highest point	**R**	**O**	**F**	**C**
29#	2″ meter	Ground	43″	X			
48#	Box of 8″ Bolts	Ground	43″				
45#	Pier block	Ground	43″		X		
50#	4′x8′ Thin Form Plywood	Ground	32″	X			
54#	Brooks #4 Box	Ground	43″	X			
60#	Tapping Motor	Ground	43″		X		
60#	Concrete/Sand	Ground	43″	X			
60#	3″ Meter	Ground	43″	X			
60#	Concrete Bag	Ground	43″	X			
60#	Hydraulic Jackhammer	Ground	43″	X			
70#	3″ Meter	Ground	43″	X			
70#	6″ Tee	Ground	43″	X			
71#	Hydraulic jackhammer	Ground	43″	X			
73″	Cart Saw	Ground	43″	X			
73″	Push bar	Ground	43″	X			
75#	4″ Valve	Ground	43″	X			
90#	Roll 2″ Copper	Ground	43″	X			
93# 46.5	Trash Pump 1-person 2-person	Ground	43″	X			
140# 70#	6″ Valve 1-person 2-person	Ground	43″	X			
130# 65#	12″ Shell/Cutter Head 1-person 2-person	Ground	43″	X			
95#	Jumping Jack Compactor	Ground	30″	X Up to 6x per day			

Courtesy of Tualatin Valley Water District

4 of 6

Figure A-23 Physical capacities requirements (heavy): maintenance and construction crew waterworks operator (*continued*)

AWWA M3 SAFETY MANAGEMENT FOR UTILITIES

Maintenance and Construction Crew Waterworks Operator

CARRYING

In an eight-hour day, this job requires: **R - Rarely** Less than 1 hr or 1–5% per day **O - Occasionally** 1–2.5 hrs or 10–33% per day 10% = 4–6×/hr, 33% = 20×/hr or < 60×/day)			**F - Frequently** 2.5–5.5 hrs or 34–66% per day (34% = > 20×/hr, 66%=120×/hr or < 720×/day) **C - Continually** 5.5 – 8 hrs or 67–100% per day (67% = > 120×/hr or > 720×/day)			
Weight	**Object**	**Maximum distance carried**	**R**	**O**	**F**	**C**
45#	Pier Block	5-10′		X		
48#	Box of 8″ Bolts	50′	X			
54#	Brooks #4 Box	20′	X			
60#	Concrete/Sand	50′	X			
60#	Tapping Motor	50′	X			
60#	Concrete Bag	10′	X			
60#	Hydraulic Jackhammer	5′-10′	X			
70#	6″ Tee	20′	X			
71#	Hydraulic Jackhammer	50′	X			
73#	Push Bar	100′		X		
73#	Cart Saw	100′		X		
75#	4″ Valve	20′	X			
90#	Roll 2″ Copper	20′	X			
93# 46.5	Trash Pump 1-person 2-person	5-10′	X			
130# 65#	12″ Shell/Cutter Head 1-person 2-person	20′	X			

Courtesy of Tualatin Valley Water District

5 of 6

Figure A-23 Physical capacities requirements (heavy): maintenance and construction crew waterworks operator (*continued*)

AWWA M3 SAFETY MANAGEMENT FOR UTILITIES

Maintenance and Construction Crew Waterworks Operator

PUSH/PULL

Weight	Object	Maximum distance	R	O	F	C
38#	Operating valves	Turning with 1 arm	X			
38# 1 arm	Valve turning-manual	Pull	X			
43#	Operating hydrants	Turning valve	X			
73#	Cart Saw	Push 100 feet		X		
80#	8" Pipe with Pry Bar	3-6" Push	X			
95#	Manhole Lid	Vertical pull upwards	X			
120#	Tightening fitting with wrench	Pull and push	X			
515# 150#	Cold Barrel Mix	Pull and tilt Push	X			

ENVIRONMENT EXPOSURES

Maintenance and Construction Crew

Inside: 20 % Outside: 80 %
Temperature Range: ______ °F to ______ °F

	Rarely	Occasionally	Frequently	Constantly	Comments
Chemicals: Contact:	x				Chlorine, marking paint, solvents, gasoline
Inhalation	x				See Chemical Contacts
Restricted Spaces		x			Confined spaces and trenches
High Elevations	n/a				
Moving Objects				x	Vehicles and equipment
Noise				x	Vehicles and equipment
Safety Equipment				x	PPE
Slippery Surface				x	
Special Clothing	n/a				
Vibration			x		Vehicles and equipment
Wetness			x		
Lighting	x				Night work
Floor Surface	n/a				

Courtesy of Tualatin Valley Water District

6 of 6

Figure A-23 Physical capacities requirements (heavy): maintenance and construction crew waterworks operator (*continued*)

Appendix B

Canadian Water Works Safety References

Following is a brief listing of regulatory federal and provincial legislation that governs safety issues in Canada and the websites where these regulations can be obtained. Also in this section is a list of websites for mandatory documents that every private and municipal water operation should have access to, with preference of a hard copy at each facility. The web links will require some independent research to assess the importance of each topic as it applies to specific Canadian jurisdictions.

Websites are listed for federal and provincial legislation and acts and the web links—current as of the publication of this manual—to access these documents. Note that the judicial system in Canada is at the top of the hierarchy for final approval, watch guarding, and enforcing safety regulations and standards.

In general, the US Occupational Safety and Health Administration workplace health and safety regulations and standards are followed almost exactly in Canada. Some differences from OSHA do exist in the Canadian federal and provincial safety standards. Canadians must comply with Canadian law with respect to safety, so it is necessary to have a good base of documentation and understanding of the Canadian Legal Acts. The links provided here will be of assistance in obtaining these critical documents for referencing, implementing programs, and providing valuable information when creating site-specific manuals regarding safety at any Canadian water works facility. The AWWA M3, *Safety Practices for Water Utilities*, should be used as a template for Canadian safety activities.

This list serves as an introduction to assist Canadian water works safety officers, joint safety committees, and management in quickly obtaining safety information for the starting a safety program and for continuing current information on safety topics as they pertain to water works operations.

CANADIAN LINKS FOR HEALTH AND SAFETY IN THE WATER INDUSTRY

Federal Code, Regulations, and Acts

At any water facility, a copy of the following four Codes, Regulations, and Acts should be in management's possession.

Canada Labour Code (R.S., 1985, c. L-2)

http://laws.justice.gc.ca/PDF/Statute/L/L-2.pdf

http://laws.justice.gc.ca/eng/L-2/index.html

Canada Occupational Health and Safety Regulations

http://laws.justice.gc.ca/PDF/Regulation/S/SOR-86-304.pdf

http://laws.justice.gc.ca/eng/SOR-86-304/index.html

Hazardous Products Act (R.S., 1985, c. H-3)

http://laws.justice.gc.ca/PDF/Statute/H/H-3.pdf

http://laws.justice.gc.ca/eng/H-3/index.html

Transportation of Dangerous Goods Act, 1992 (1992, c. 34)

http://laws.justice.gc.ca/PDF/Statute/T/T-19.01.pdf

http://laws.justice.gc.ca/eng/T-19.01/index.html

Additional important federal links

Workplace Hazardous Materials Information System–Official National Site

http://www.hc-sc.gc.ca/ewh-semt/occup-travail/whmis-simdut/index-eng.php

Environmental & Workplace Health – Water Quality – Drinking Water

http://www.hc-sc.gc.ca/ewh-semt/water-eau/drink-potab/index-eng.php

Environmental and Workplace Health

http://www.hc-sc.gc.ca/ewh-semt/occup-travail/index-eng.php

Environmental Scan on Workplace Health in Canada (2009)

http://www.healthyworkenvironments.ca/announcements/environmental-scan-on-workplace-health-in-Canada

Government of Canada Health for Industry

http://canada.ca/en/services/health/industries.html

Provincial Codes, Regulations, and Acts

Alberta

Occupational Health and Safety Act

http://www.qp.alberta.ca/574.cfm?page=O02.cfm&leg_type=Acts&isbncln=0779749200

Occupational Health and Safety Regulation

http://www.qp.alberta.ca/574.cfm?page=2003_062.cfm&leg_type=Regs&isbncln=077971752X

British Columbia

Workers Compensation Act

http://www.bclaws.ca/EPLibraries/bclaws_new/document/ID/freeside/96492_02

Occupational Health and Safety Regulation

http://www.bclaws.ca/EPLibraries/bclaws_new/document/ID/freeside/296_97_01

Manitoba

The Workplace Safety and Health Act

http://web2.gov.mb.ca/laws/statutes/ccsm/w210e.php

New Brunswick

Occupational Health and Safety Act

http://www.gnb.ca/0062/acts/acts/o-00-2.htm

Occupational Health and Safety Act Regulations

http://www.gnb.ca/0062/regs/o-0-2reg.htm

Newfoundland and Labrador

An Act Respecting Occupational Health and Safety in the Province

http://www.assembly.nl.ca/Legislation/sr/statutes/o03.htm

Newfoundland and Labrador Regulation 70/09

http://www.assembly.nl.ca/Legislation/sr/Regulations/rc090070.htm

Nova Scotia

Occupational Health and Safety Act

http://www.gov.ns.ca/legislature/legc/statutes/occph_s.htm

Occupational Safety General Regulations

http://www.gov.ns.ca/just/regulations/regs/ohsgensf.htm

Northwest Territories

Mine Health and Safety Act

http://www.justice.gov.nt.ca/Legislation/..%5CPDF%5CREGS%5CMINE_HEALTH_ SAFETY/Mine_Health_and_Safety.pdf

Nunavut

Consolidation of Mine Health and Safety Regulations

http://www.canlii.org/en/nu/laws/regu/nwt-reg-nu-125-95/latest/part-1/nwt-reg-nu-125-95-part-1.pdf

Ontario

Occupational Health and Safety Act

http://www.e-laws.gov.on.ca/html/statutes/english/elaws_statutes_90o01_e.htm

Part X Regulations

http://www.e-laws.gov.on.ca/html/statutes/english/elaws_statutes_90o01_e.htm#BK79

Prince Edward Island

Occupational Health and Safety Act

http://www.gov.pe.ca/law/statutes/pdf/o-01_01.pdf

Quebec

An Act Respecting Occupational Health and Safety

http://www.canlii.org/en/qc/laws/stat/rsq-c-s-2.1/latest/rsq-c-s-2.1.html

Saskatchewan

The Occupational Health and Safety Act, 1993

http://www.qp.gov.sk.ca/documents/English/Statutes/Statutes/O1-1.pdf

The Occupational Health and Safety Regulations, 1996

http://www.qp.gov.sk.ca/documents/English/Regulations/Regulations/O1-1R1.pdf

Yukon

O.I.C. 2006/178 Occupational Health and Safety Act

http://www.gov.yk.ca/legislation/regs/oic2006_178.pdf

Additional Websites for Provinces and Territories

Federal page with links to the official government websites of Canada's provinces and territories

http://www.pco-bcp.gc.ca/aia/index.asp?lang=eng&page=relations&doc=offices/ offices-eng.htm

The Occupational Health Clinics for Ontario Workers Inc.

http://www.ohcow.on.ca/

Links to Workers' Compensation Boards/Commissions

http://www.awcbc.org/en/linkstoworkerscompensationboardscommissions.asp

Acronyms

AED	automated external defibrillator
AFH	arc flash hazards
AIHA	American Industrial Hygiene Association
ANSI	American National Standards Institute
ASME	American Society of Mechanical Engineers
ASSE	American Society of Safety Engineers
BPVC	(ASME) Boiler and Pressure Vessel Code
CCPS	Center for Chemical Process Safety
CCR	Consumer Confidence Report
CDC	Centers for Disease Control and Prevention
CFR	Code of Federal Regulations
CPR	cardiopulmonary resuscitation
dBA	decibel level
DEET	*N,N*-Diethyl-*meta*-toluamide
DMV	Department of Motor Vehicles
EMR	experience modification rating
EMS	emergency medical services
EPA	employee assistance program
ERTW	early return to work
GHS	globally harmonized system
HASP	health and safety plan
HPS	hantavirus pulmonary syndrome
HSE	health, safety, and environmental
HIV	human immunodeficiency virus
IIPP	injury and illness prevention programs
IIR	incident investigation report
ITCP	internal traffic control plan
IT	information technology
JHA	job hazard analysis
JSA	job safety analysis
LWD	lost workday
MP	management programs
MSDs	musculoskeletal disorders
MSDS	material safety data sheets
NA	not applicable
NASA	National Aeronautics and Space Administration
NCADD	National Council on Alcoholism and Drug Dependence, Inc.
NFPA	National Fire Protection Association
NIOSH	National Institute for Occupational Safety and Health

O&M	operations and maintenance
OCIR	Office of Community & Intergovernmental Relations
OHP	occupational health provider
OHSAS	Occupational Health and Safety Assessment Series
OSH	Occupational Safety and Health
OSHA	Occupational Safety and Health Administration
PAPR	powered air-purifying respirator
PCT	physical capacity test
PPE	personal protective equipment
PSI	per square inch
PSM	process safety management
PtD	prevention through design
PTO	power take off
RMP	risk management program
ROPS	rollover protective structures
RTW	return to work
SCA	safety and compliance assistance
SCA	sudden cardiac arrest
SCBA	self-contained breathing apparatus
SDS	safety data sheet
TB	tuberculosis
TCP	traffic control plan
TWA	time-weighted average
USEPA	US Environmental Protection Agency
WRF	Water Research Foundation
WTPs	water treatment plants
XRF	x-ray fluorescence

Index

Note: *f.* indicates figure; *t.* indicates table

A

accident/incident analysis, 119
accident investigation program, 116
administrative controls, 28
American National Standards Institute, 4
and Illness Prevention Programs (IIPP), 4
ANSI/AIHA standard Z10-2012, 5
asbestos, 101
audits, 18
AWWA Water Sector Utility Health, Safety and Environmental (HSE) Program Survey, 43

B

basic steps, writing the JSA
  list job tasks, 24
  look for hazards, 24
  recommended controls and actions, 27
biological hazards, 95
bloodborne pathogens, 97
buddy system, 114

C

Canadian water works safety references, 155
cardiopulmonary resuscitation (CPR), 32
Certificate of Recognition (COR), 2
chemical hazards, 98
common controls, 88–89, 102
common hazards and safety programs, 77
compressed gases, 99
conservation coordinator, 30
contractor acceptance criteria, 60
contractor evaluation, 64
contractor prequalification, 58
contractors, regulatory requirements and best practices, 57
contractor safety management, 53
contractor work rules, 62
contract requirements, 58
control elements, 104
corporate policies and programs, 9
corrective action, 16, 38
corrective and preventive action, 18
cost–benefit analysis, 38

D

day-to-day activities and priorities, 6
design methodology, 27
developing written programs
  compliance programs, 13
  employee participation/awareness, 12
  employee training, 12
  HSE responsibilities, 12
  implementation matrix, 14
  injury/incident investigation and reporting, 12
  know your audience, 13
  know your objectives, 13
  management program compliance, 12
  management program development, 12
  procedure development, 13
  program improvement (auditing), 13
  use a consistent format, 14
disease prevention, 112
driving hazards, 91
duties, safety committee
  accident investigations, 8
  developing safe work practices, 8
  developing written safety programs, 8
  facilitating safety training, 8
  workplace self-inspections, 8

E

early return to work program, 123*f*
employee conflict, 113
employee participation, 4, 5
employees, safety and health, 1
engineered controls, 27
environmental program, 1
environment exposures, 154
equipment specifications, 49
ergonomic hazards, 108

essential job functions, 120
excavations, 80

F

feedback to planning, 18

G

general utility HSE responsibilities, 11*t*
good hazard scenarios, 25
ground fault interrupter, 82*f*
guidance documents, 115

H

hazard action list, 30
hazard communication, 1
hazard identification, 48
hazardous energy, release, 53
hazardous energy control (electrical safety), 81
hazardous waste, 105
hazard summary/action list, 127*f*
Health, Safety, and Environmental (HSE) program, 1
health and safety in the water industry, Canadian links, 156
hot and cold weather conditions, 93
hot weather hazards, 93
hot work, 83
hot work permit, 128*f*
HSE management programs, 4
HSE management systems, 17

I

important safety factor, JSA, 29
incident investigation, 17
incident management, 31
incident management programs, 31
incident reporting, 33
incident reporting and investigation, 63
incident weed and root cause analysis, 37*f*
initial medical attention, 32
injury/illness recordkeeping and reporting, 11
internal design standards, 43
interviewing witnesses, 34
investigation and documentation, 34

J

job description, 30
job description (heavy physical requirements), 149*f*
job safety analysis, 21, 30, 130*f*
job safety analysis form, 129
JSA, advantages, 22
JSA, preparing, 23
JSA, purpose, 26
JSA development, 23

L

large utility safety organization, example, 10*f*
lone worker, 110

M

maintenance and construction crew, 154
maintenance and construction crew waterworks operator, 152*f*
major injury, 117
management commitment, 5, 7
management leadership, 4, 5
materials handling, 84
medical assistance, 109
minimum requirements, 121

N

National Institute for Occupational Safety and Health (NIOSH), 43
noise exposure, 85
non-work related injury/illness, 124

O

observable management involvement, 7
Occupational Health and Safety, 4
occupational health provider (OHP), 32
Occupational Safety and Health Administration (OSHA), 21
O&M professionals, roles and responsibilities, 46
operations and maintenance (O&M), 43
operations and maintenance (O&M) staff, 27
organizational culture, 6
OSHA quick card, 137*f*

P

passing criteria, 133*f*
permit-required confined space entry, 79
personal (or health) hazards, 107
personal protective equipment (PPE), 28
pesticides (licensed applicators), 106
physical capacities
    requirements, 122*f*, 134*f*, 151*f*
physical harm, 2
physical hazards, 78
policies and management plans, 6
post-incident response, 33
pre-job contractor orientation, 61
pre-placement physical capacities test, 136*f*
prequalification, examples, 59*t*
pressure vessels, 86
prevention methods, inclusion, 44
Prevention through Design [PtD], 43
proactive design safety review, 50
Process Safety Management (PSM), 4
program staffing, 9
project management, 47
properly stacking, 100
protection, common form, 28
protective procedures, 26
PTD implementation, 45
PTD initiative, 44
PTD methodology, 48
Public Utility Commission (PUC), 56

R

rail fall protection system, 44
record keeping, 16, 40
release to return to work, 135*f*
respiratory hazards, 111
return to full duty, 126
return to work, 125
return-to-work programs, 39
return-to-work programs, key aspects
  claims management, 40
  identification of light duty, 39
  knowledge of physical requirements, 39
  management support, 39
  occupational health provider, 39
return- to-work (RTW) programs, 32
roles and responsibilities, 10
root cause analysis, 35

S

safety culture, 7
safety data sheet sample, 138*f*
safety data sheet (SDS), 26
safety program elements, 11
safety staffing, 9
sample job safety analysis, 131*f*
silica, 103
sketching motion through time, 36*f*

T

temporary modified duty tasks, 148*f*
Texas Department of Insurance, Workers
    Compensation Division, 3
Texas public worker health and safety, 3
Texas Workers' Compensation Act, 3
third-party contractor safety
    management, 65
tractor operations, 87
traffic and the work zone, 90
training and education, 14
training documentation, 15
treatment chemicals, 45
triangulation method, 35*f*
type of operation, percent of lost
    workdays, 55*f*

U

US Environmental Protection Agency
    (USEPA) Risk Management Program
    (RMP), 4
US Occupational Safety and
    Health Act of 1970, 2
US Occupational Safety and Health
    Administration workplace health
    and safety regulations, 155

V

vehicle accident, 118
video recordings, 34
violence in the workplace, 112

W

waterborne pathogens, 96
Water Research Foundation (WRF)
    project, 31
water transmission, 45
5-Why technique, 36
work environment, 43
work-related injury/illness, 123

This page intentionally blank.

AWWA Manuals

M1, *Principles of Water Rates, Fees, and Charges*, #30001

M2, *Instrumentation and Control*, #30002

M3, *Safety Management for Water Utilities*, #30003

M4, *Water Fluoridation Principles and Practices*, #30004

M5, *Water Utility Management*, #30005

M6, *Water Meters—Selection, Installation, Testing, and Maintenance*, #30006

M7, *Problem Organisms in Water: Identification and Treatment*, #30007

M9, *Concrete Pressure Pipe*, #30009

M11, *Steel Pipe—A Guide for Design and Installation*, #30011

M12, *Simplified Procedures for Water Examination*, #30012

M14, *Backflow Prevention and Cross-Connection Control: Recommended Practices*, #30014

M17, *Fire Hydrants: Installation, Field Testing, and Maintenance*, #30017

M19, *Emergency Planning for Water and Wastewater Utilities*, #30019

M20, *Water Chlorination/Chloramination Practices and Principles*, #30020

M21, *Groundwater*, #30021

M22, *Sizing Water Service Lines and Meters*, #30022

M23, *PVC Pipe—Design and Installation*, #30023

M24, *Planning for the Distribution of Reclaimed Water*, #30024

M25, *Flexible-Membrane Covers and Linings for Potable-Water Reservoirs*, #30025

M27, *External Corrosion Control for Infrastructure Sustainability*, #30027

M28, *Rehabilitation of Water Mains*, #30028

M29, *Water Utility Capital Financing*, #30029

M30, *Precoat Filtration*, #30030

M31, *Distribution System Requirements for Fire Protection*, #30031

M32, *Computer Modeling of Water Distribution Systems*, #30032

M33, *Flowmeters in Water Supply*, #30033

M36, *Water Audits and Loss Control Programs*, #30036

M37, *Operational Control of Coagulation and Filtration Processes*, #30037

M38, *Electrodialysis and Electrodialysis Reversal*, #30038

M41, *Ductile-Iron Pipe and Fittings*, #30041

M42, *Steel Water-Storage Tanks*, #30042

M44, *Distribution Valves: Selection, Installation, Field Testing, and Maintenance*, #30044

M45, *Fiberglass Pipe Design*, #30045

M46, *Reverse Osmosis and Nanofiltration*, #30046

M47, *Capital Project Delivery*, #30047

M48, *Waterborne Pathogens*, #30048

M49, *Quarter-Turn Valves: Head Loss, Torque, and Cavitation Analysis*, #30049

M50, *Water Resources Planning*, #30050

M51, *Air Valves: Air-Release, Air/Vacuum and Combination*, #30051

M52, *Water Conservation Programs—A Planning Manual*, #30052

M53, *Microfiltration and Ultrafiltration Membranes for Drinking Water*, #30053

M54, *Developing Rates for Small Systems*, #30054

M55, *PE Pipe—Design and Installation,* #30055

M56, *Nitrification Prevention and Control in Drinking Water,* #30056

M57, *Algae: Source to Treatment,* #30057

M58, *Internal Corrosion Control in Water Distribution Systems,* #30058

M60, *Drought Preparedness and Response,* #30060

M61, *Desalination of Seawater,* #30061

M62, *Membrane Applications for Water Reuse,* #30062

M63, *Aquifer Storage and Recovery,* #30063

M65, *On-Site Generation of Hypochlorite,* #30065

M66, *Cylinder and Vane Actuators and Controls—Design and Installation,* #30066

M68, *Water Quality in Distribution Systems,* #30068

M69, *Inland Desalination and Concentrate Management,* #30069

M77, *Condition Assessment of Water Mains,* #30077

www.ingramcontent.com/pod-product-compliance
Lightning Source LLC
LaVergne TN
LVHW082337200926
848481LV00012B/3368
* 9 7 8 1 6 4 7 1 7 0 7 7 6 *